About IFPRI

The International Food Policy Research Institute (IFPRI®) was established in 1975 to identify and analyze alternative national and international strategies and policies for meeting food needs of the developing world on a sustainable basis, with particular emphasis on low-income countries and on the poorer groups in those countries. While the research effort is geared to the precise objective of contributing to the reduction of hunger and malnutrition, the factors involved are many and wide-ranging, requiring analysis of underlying processes and extending beyond a narrowly defined food sector. The Institute's research program reflects worldwide collaboration with governments and private and public institutions interested in increasing food production and improving the equity of its distribution. Research results are disseminated to policymakers, opinion formers, administrators, policy analysts, researchers, and others concerned with national and international food and agricultural policy.

About IFPRI Food Security in Practice series

The Food Security in Practice technical guide series is designed for development practitioners. The guides are based on IFPRI research and enable project personnel in the field to take research from analysis to action. Each volume addresses informational and methodological issues that practitioners confront during the life of a project and presents the lessons learned from research on specific development issues. Relevant research and operational concepts are explained in easy-to-understand ways. Additional information pertaining to research analyses, methodologies, and results is available from IFPRI.

FOOD SECURITY IN PRACTICE

Strategic Analysis and Knowledge Support Systems for Agriculture and Rural Development in Africa

Translating Evidence into Action

Michael Johnson and Kathleen Flaherty

International Food Policy Research Institute
2033 K Street, N.W.
Washington, D.C. 20006-1002
U.S.A.
Telephone +1-202-862-5600
www.ifpri.org

How to cite this book: Johnson, M., and K. Flaherty. 2011. *Strategic analysis and knowledge support systems for agriculture and rural development in Africa: Translating evidence into action.* Food Security in Practice technical guide 6. Washington, D.C.: International Food Policy Research Institute.

DOI: 10.2499/9780896297845fsp6

Library of Congress Cataloging-in-Publication Data

Johnson, Michael (Michael Emmett), 1963–
Strategic analysis and knowledge support systems for agriculture and rural development in Africa : translating evidence into action / by Michael Johnson and Kathleen Flaherty.
p. cm. — (Food security in practice ; no. 6)
Includes bibliographical references.
ISBN 978-0-89629-784-5 (alk. paper)
1. Agricultural development projects—Africa, Sub-Saharan. 2. Rural development—Africa, Sub-Saharan. 3. Strategic planning—Africa, Sub-Saharan. I. Flaherty, Kathleen. II. Title. III. Series: Food security in practice technical guide series ; no. 6.
HD2117.J64 2011
338.1'8670684—dc22 2010034471

Contents

Tables, Figures, and Boxes

Tables

Figures

Boxes

Preface: A Brief History of SAKSS

The Strategic Analysis and Knowledge Support System (SAKSS) is defined as a network of people and institutions that provides timely, credible, and evidence-based knowledge and analysis to inform agricultural and rural development (ARD) strategies in Africa.[1] It was conceived in 2003 by researchers at the International Food Policy Research Institute (IFPRI) based on its many years of providing key data analysis, policy research, and capacity strengthening to governments and donors in Africa as they formulated and implemented their development strategies. The network involves a number of key international and local actors from academic, research, government, and nongovernmental institutions located in a specific country or region, including researchers, policy analysts, statisticians, geographic information system specialists, government technocrats, practitioners, and policymakers. The actors who make up the SAKSS network are relied on to contribute their expertise and knowledge to the analysis and peer review of the evidence generated for ARD strategies and to the dialogue surrounding it.

Originally the SAKSS network was set up to support the design and implementation of the U.S. Presidential Initiative to End Hunger in Africa (IEHA) for the United States Agency for International Development. Its activities involved undertaking strategic analysis designed to fill knowledge gaps quickly, prioritize future investments, and provide guidance to IEHA for monitoring and evaluation of impact (Johnson et al. 2003).

One of the goals of SAKSS, as conceived, is to promote dialogue among research, practitioner, and policymaking communities in ARD in order to better articulate the type of development strategy questions being asked, the analysis and corresponding information needed, and the results to help guide and monitor investments. The involvement of local partners in setting up a SAKSS for IEHA ultimately provided broader exposure to many more potential users and suppliers of strategic analysis and information for ARD strategies. It did not take long, therefore, for the SAKSS concept to begin generating interest among the African and donor communities concerned with ARD. The launching of a joint priority-setting study by the Association for Strengthening Agricultural Research in Eastern and Central Africa and IFPRI in 2004 played a key role by illustrating how strategic analysis (as a key component of SAKSS) can contribute to the priority-setting exercises of development strategy. The study's final report came out at a time when African governments were increasingly being challenged, especially by the poverty reduction strategy paper process, to show evidence of how their chosen strategies would lead to growth and poverty reduction.[2]

The demands for greater evidence-based decisionmaking in Africa also occurred during a period when IFPRI was undergoing some structural changes of its own. In addition to the creation of a new research division on development strategies and governance, the institution moved toward greater decentralization by setting up country and regional program offices in Africa, Asia, and Latin America. Within Africa, the application of the SAKSS concept quickly proved a constructive framework for launching new IFPRI country programs—later referred to as country strategy support programs (CSSPs). The CSSPs in Ethiopia, Ghana, Nigeria, and Uganda have since expanded on the original SAKSS concept to include a broader range of analyses and knowledge systems approaches based on the needs and capacities of their network of national partner institutions and individuals.

At the multicountry level, the SAKSS concept was also adopted as a framework to help establish an Africawide network in support of the implementation of the Comprehensive Africa Agriculture Development Programme (CAADP) of the African Union (AU) and the New Partnership for Africa's Development (NEPAD), referred to as the Regional Strategic Analysis and Knowledge Support System. Three nodes were set up in each of three African regional economic communities (the Common Market of East and Southern Africa, the Southern African Development Community, and the Economic Community of West African States) under the guidance and cooperation of IFPRI and four other Africa-based centers of the Consultative Group on International Agricultural Research (the International Livestock Research Institute [ILRI], the International Institute of Tropical Agriculture, the International Crops Research Institute for the Semi-Arid Tropics, and the International Water Management Institute). Each node represents a broad network of national and regional organizations and individuals that supply data, analysis, and research according to a common work plan developed in consultation with the regional economic communities, NEPAD, and the AU. Members from each country include individuals from analytical units of agricultural ministries, statisticians, university professors, researchers from national agricultural research institutions and think tanks, and others.

Today, both the country and regional SAKSS networks share the principles of (1) providing timely and credible

[1] See Johnson and Resnick (2004).

[2] This study was later published as Omamo et al. (2006).

analysis and data to policymakers to strengthen the evidence during deliberations about future ARD investments and policies, (2) promoting locally relevant research and analysis based on need, and (3) working to strengthen local capacities for analysis and evidence-based dialogue.

As the demands for the establishment of a country SAKSS, especially in support of CAADP implementation, continue to grow in many African countries, the need for IFPRI to prepare a technical guide on SAKSS has become critical. This technical guide primarily builds on the experience of IFPRI and its establishment in Africa of country and regional SAKSSs, which have now been in operation for several years. These applications should offer a real world opportunity to draw comparative lessons that can help guide future efforts in establishing similar networks elsewhere in Africa as the need arises.

Acknowledgments

The authors acknowledge the helpful comments of a number of internal and external reviewers during the preparation of this technical guide. The contribution of many other researchers within and outside of the International Food Policy Research Institute (IFPRI) during the initial inception and later refinement of the Strategic Analysis and Knowledge Support System concept is well noted. In particular, the early thinking of Peter Hazell (while still serving as a division director at IFPRI), Jeff Hill (United States Agency for International Development, Bureau for Africa), and Stanley Wood (IFPRI) shaped many of the ideas inherent in this concept for Africa. The authors are indebted to them. In later years, Shenggen Fan (IFPRI) and the entire Development Strategy and Governance Division at IFPRI helped refine the Strategic Analysis and Knowledge Support System concept as it moved from being only an idea to being incorporated into many of the Division's country support programs and into the support provided to the Comprehensive Africa Agriculture Development Programme of the African Union's New Partnership for Africa's Development. The assistance and earlier writings of Danielle Resnick and Tsitsi Makombe are also acknowledged.

Acronyms and Abbreviations

AIC	Akaike's information criteria
APSF	Agricultural Policy Support Facility (Nigeria)
ARD	agricultural and rural development
ASARECA	Association for Strengthening Agricultural Research in Eastern and Central Africa
AU	African Union
CAADP	Comprehensive Africa Agriculture Development Programme
CGE	computable general equilibrium
CGIAR	Consultative Group on International Agricultural Research
CIDA	Canadian International Development Agency
COMESA	Common Market for Eastern and Southern Africa
CORAF/WECARD	West and Central African Council for Agricultural Research and Development
CSSP	country strategy support program
DFID	Department for International Development (U.K.)
DREAM	Dynamic Research EvaluAtion for Management
DSGD	Development Strategy and Governance Division
ECA	East and Central Africa
ECOWAS	Economic Community of West African States
EDRI	Ethiopian Development Research Institute
EMM	economywide multimarket model
ESSP	Ethiopia Strategy Support Program
FAO	Food and Agriculture Organization of the United Nations
FARA	Forum for Agricultural Research in Africa
FASDEP	Food and Agricultural Sector Development Policy
FMAWR	Federal Ministry of Agriculture and Water Resources
GIS	geographic information systems
GSSP	Ghana Strategy Support Program
ICRISAT	International Crops Research Institute for the Semi-Arid Tropics
ICT	information and communication technology
IDRC	International Development Research Centre
IEHA	Initiative to End Hunger in Africa
IFDC	International Center for Soil Fertility and Agricultural Development
IFPRI	International Food Policy Research Institute
IITA	International Institute of Tropical Agriculture
ILRI	International Livestock Research Institute
IMPACT	International Model for Policy Analysis of Agricultural Commodities and Trade
IWMI	International Water Management Institute
M&E	monitoring and evaluation
MAPS	Marrakech Action Plan for Statistics
MDG	Millennium Development Goal
MoFA	Ministry of Food and Agriculture
MoFEP	Ministry of Finance and Economic Planning
NAC	National Advisory Committee
NARI	National Agricultural Research Institute
NARS	National Agricultural Research System
NEPAD	New Partnership for Africa's Development
NGO	nongovernmental organization

ODI	Overseas Development Institute
OECD	Organization for Economic Co-operation and Development
PMA	Plan for the Modernization of Agriculture
PRS	poverty reduction strategy
PRSP	poverty reduction strategy paper
R&D	research and development
REC	regional economic community
REKSS	rural economy knowledge support system
ReSAKSS	Regional Strategic Analysis and Knowledge Support System
ReSAKSS-ECA	ReSAKSS–East and Central Africa
ReSAKSS-SA	ReSAKSS–Southern Africa
SADC	Southern African Development Community
SAKSS	Strategic Analysis and Knowledge Support System
SAM	social accounting matrix
SCRIP	Strategic Criteria for Rural Investments in Productivity
Sida	Swedish International Development Cooperation Agency
USAID	United States Agency for International Development
USSP	Uganda Strategy Support Program
WFP	World Food Programme (UN)
WSM	Water Simulation Model

Summary

In this technical guide we describe the Strategic Analysis and Knowledge Support System (SAKSS) concept developed by the International Food Policy Research Institute in response to the capacity gaps typically found in many African countries for generating and translating evidence into action. SAKSS is defined as a collaborative network that acts as a mechanism by which relevant evidence is generated and used to inform agricultural and rural development strategy formulation and implementation. SAKSS brings together two important concepts—"strategic analysis" and "knowledge support systems." The *strategic analysis* component describes an integrated framework of analysis that helps identify strategy and investment options for achieving high-end development goals. This analysis is carried out within the context of a *knowledge support system,* a network that serves the evidence needs of strategy formulation and implementation. By providing a platform for exchange, information, data analysis, and knowledge can be compiled, synthesized, and packaged into evidence that is supplied on a timely and reliable basis to be of use during strategy processes.

The purpose of the technical guide is to present a useful and practical guideline for applying the SAKSS concept and establishing a country SAKSS based on the experience to date. The need for this has come about due to a growing demand for the kinds of applications offered by the SAKSS as many more countries, especially in Sub-Saharan Africa, express their wish to establish similar knowledge support systems. The technical guide is therefore intended to serve policy analysts and researchers who are helping to set up country SAKSSs, African government and development partners who wish to use SAKSS for their evidence needs, and others who wish to participate in such a network. It is organized in three main parts. The first part reviews the SAKSS concept and approach to provide a background definition of its objectives and underlying principles. The second part provides a practical guide on how to set up a country SAKSS, followed by examples of existing applications. The final part serves as an appendix of tools and resources.

Strategic Analysis and Knowledge Support Systems for Agriculture and Rural Development in Africa

Translating Evidence into Action

Chapter 1: Introduction and Overview

More than 1 billion people worldwide are living in extreme poverty—meaning that they earn less than US$1 per day—and many suffer from hunger and malnutrition. In developing countries, about 25 percent of children under the age of five are malnourished. Encouragingly, many developing countries have committed to the Millennium Development Goals (MDGs), which include halving poverty and hunger by 2015 (MDG1). In Sub-Saharan Africa, the problem is particularly acute, and policymakers have been called on to allocate more resources and design strategies to accelerate agricultural growth in order to meet MDG1. Their efforts are reflected in the poverty reduction and agricultural and rural development strategies being formulated by individual countries.

THE IMPORTANCE OF AGRICULTURE AND RURAL DEVELOPMENT STRATEGIES

The move toward more poverty-focused strategies in Africa at the turn of the century helped draw attention to agriculture and rural development (ARD) and the process by which the strategies are formed, implemented, and evaluated. Beginning with the poverty reduction strategy papers initially introduced by the World Bank and the International Monetary Fund as a means by which countries would access concessional lending and debt relief under the Enhanced Highly Indebted Poor Country Initiative, there was a push toward encouraging national ownership and wider participation in the process of defining priorities. This process forced planners to recognize the importance of rural poverty reduction. It also introduced debates on the question of which productive sectors contribute the most to pro-poor growth, including the role of the state (Cabral 2006).

More recently, the launching of a continentwide and Africa-led Comprehensive Africa Agriculture Development Programme (CAADP) of the African Union and New Partnership for Africa's Development (NEPAD) has drawn further attention to ARD. Though CAADP, African governments in 2003 agreed to the shared commitment of allocating at least 10 percent of their national budgets to agriculture in order to achieve a 6 percent annual sector growth rate and meet MDG1, halving poverty by 2015. Today a number of African countries have met or surpassed this goal, including Chad, Ethiopia, Madagascar, Mali, Namibia, and Niger, with others not very far behind, including Kenya, Malawi, Rwanda, and Zambia (Fan, Omilola, and Lambert 2009).

CAADP identifies four priorities or "pillars" that require investments and action: extending the area under sustainable land management and reliable water control systems (for example, by increasing access to irrigation), increasing market access through improved rural infrastructure and other trade-related interventions, increasing food supply and reducing hunger across the region by increasing smallholder productivity and improving responses to food emergencies, and improving agricultural research and systems to disseminate appropriate new technologies and increasing the support given to farmers to adopt them (NEPAD 2003). As a continentwide initiative, CAADP has played a particularly important role in encouraging greater donor harmonization surrounding their investments in African ARD.[3] At the regional and local levels, CAADP is also providing a framework by which countries can harmonize their policies to stimulate agriculture growth and poverty reduction on the continent. The challenge now is to ensure that the selected priorities of the ARD strategies under these various initiatives will produce their intended goals of sustaining broad-based growth and poverty reduction on the continent.

Finally, the focus on agricultural development in the *World Development Report 2008* (World Bank 2007b) exemplifies this shifting paradigm among donors and African governments. The report provides a useful synopsis of the stock of evidence and knowledge supporting the arguments for focusing more attention on ARD in Africa.

THE NEED TO STRENGTHEN KNOWLEDGE SYSTEMS

The push for effective ARD strategies to meet MDG1 has revived questions about the process of creating and implementing those strategies. Many past studies have documented the failure of planning for implementation (Wildavsky 1973; Killick 1976; Montjoy and O'Toole 1979; Brinkerhoff 1996; Crosby 1996). In particular, the inadequacies of central planning in the past have led to calls for strategy formulation that is evidence based, decentralized, participatory, and accompanied by integrated monitoring and evaluation (M&E) systems under the assumption that such mechanisms improve the implementation and outcomes of a strategy. Strategies are now seen as "living"

---

[3] See <www.donorplatform.org> (accessed July 28, 2009).

documents that allow adaptation to changing conditions in recognition of the nonlinear and dynamic nature of the strategy design and implementation process. Many different variables can affect how well a strategy is implemented, including leadership, participation, prioritization, timing, and degree of organizational and process integration (Gijsbers et al. 2001). In addition to these factors, the availability of political, financial, and technical resources to implement the strategy must be taken into account during the strategy design process (Thomas and Grindle 1990).

The complexity of formulating and implementing effective strategies demands a lot of evidence to allow a government to accurately assess the choices available to it and the trade-offs inherent in any choice they make. By *evidence* we mean data statistics and analysis of past trends, economic analysis of future policy alternatives, impact assessment of past investments, research findings from a number of disciplines (public policy, socioeconomic and political sciences, and biophysical sciences), and lessons from practice and experience. To supply this evidence, the country must have a solid foundation of analytical capacity throughout its planning agencies and academic institutions. In addition, governments need policymakers who have the motivation and ability to demand and use the information (Omamo 2004). How well national knowledge systems can bridge the supply of and demand for evidence is therefore critical for success.

National knowledge system describes the existing stock of knowledge and established links between people and organizations in influencing the type of knowledge generated and used as input in policy dialogue and decisionmaking. In this regard *knowledge systems* describes a combination of people (researchers, policy analysts, development practitioners, decisionmakers, and other stakeholders), organizations (policy analysis think tanks, universities, government agencies, nongovernmental organizations [NGOs], and private-sector groups), and the knowledge that is accumulated and shared among them (results of research and analysis, information on synthesis and trends, and local knowledge and expertise).[4] How effective the knowledge system is in generating and promoting the use of evidence will depend on many factors: the perceived credibility and relevance of the evidence generated, the type of relationships and linkages that develop among all individual actors and organizations involved, capacities to comprehend and use the evidence, the local policy process and political climate, and local beliefs and norms (Cash et al. 2003; Young 2005). Often it is the linkages among individuals and organizations that help to bridge the gap between the supply of and demand for evidence that are especially weak. In such cases it is important to establish effective mechanisms by which both sides can be more closely tied as part of ongoing dialogue and decisionmaking processes. These mechanisms effectively link suppliers and users of knowledge through the creation and use of knowledge products.

The poor state of knowledge systems for informing strategy is a serious problem in many African countries. Data collection and analysis suffer from a shortage of attention and resources. Knowledge sharing is often minimal, with planning ministries that operate in isolation as well as uncoordinated ministries, research institutes, and statistical bureaus. Government agencies, NGOs, and development partners carry out parallel and overlapping processes of information gathering. Often development partners have more input into the strategy process than do legislative bodies or the national civil society. The M&E frameworks of many strategies rarely deal with issues of causality and attribution among investments, policy changes, and outcomes.

Recent studies on bridging research and policy highlight many of these gaps in Africa and how they can be improved. Important factors include the relevance and credibility of the research, the extent of external influences (for example, donors), and the type of knowledge systems in place to help link the researchers and policymakers (Young 2005). These links, categorized as feedback, dialogue, and collaboration, are the basis for communication and involve perceptions of trust, legitimacy, and participation (Cash et al. 2003). Networks and policy communities, both formal and informal, have been found to improve such linkages (Creech and Willard 2001). In particular, feedback loops among the stakeholders and throughout the design, implementation, and monitoring processes can help to identify problems and allow adjustments to be made if needed. However, improving linkages is rarely that straightforward. The political context and the organizational and individual relationships involved all influence the extent to which research ultimately plays a role in ARD strategy formulation and implementation (Young 2005). For example, more open political systems and processes and strong demand from policymakers can increase the use of research, whereas opposing vested interests can limit its impact. Meanwhile, knowledge itself is not value neutral, because it reflects the power structure, discourse, and narratives of the institutions in which it is created.

Although bridging the research and policy divide can seem a complex and daunting task, the large capacity gaps in Africa add another layer of challenges that only exacerbate the already weak links between the supply of and demand for research. For example, local universities rarely undertake research directly relevant to local decisionmaking needs, and national institutions and agencies seldom have sufficient capacities and experience to provide relevant information

[4] See Cash et al. (2003) for a useful definition of *knowledge systems* for sustainable development.

needed to guide strategy formulation and implementation. The Strategic Analysis and Knowledge Support System (SAKSS) concept has therefore been developed in direct response to these serious capacity gaps typically found in many African countries.

Introducing SAKSS

SAKSS is a concept that defines a network of individuals and institutions that are linked in ways intended to help bring strategic analysis and research evidence to bear during the design and implementation of ARD strategies in Africa. The evidence is considered strategic because it describes information that is not only scientifically credible but has important relevance to the range of questions and issues being faced by decisionmakers as they formulate and implement ARD strategies. In particular, SAKSS brings together two important additional concepts—"strategic analysis" and "knowledge support systems"—which define its overall purpose and utility for supporting ARD in Africa.

The strategic analysis component

Strategic analysis describes a logical series of analyses that help identify policy and investment options for achieving growth and poverty outcomes, beginning with a broader, economywide perspective and ending with a more targeted sector- and community-level perspective. This type of integrated analysis is intended to guide a credible action plan of development priorities in ARD that contribute the most to the achievement of desirable targets for growth and poverty reduction (see Figure 2.1). Such prioritization implies finding answers to a range of strategic questions: What is the role of agriculture in promoting overall economic growth and poverty reduction in the different stages of development given a country's natural resource endowments? How should public resources be mobilized and allocated among different sectors, subsectors, and regions? What have been the lessons and effect of ARD strategies on outcomes and impact? Answers to these questions can help arm policymakers with useful evidence on the kinds of trade-offs and outcomes associated with their policy and investment choices. For example, although high-value commodities such as horticulture offer higher returns in terms of farm income, they do not have a broad-based effect on overall growth and poverty reduction relative to food staples (Diao et al. 2007a). This is because few can afford to participate except for those in proximity to large urban centers or ports.

Boxes 1.1 and 1.2 illustrate some of the strategic analysis that has been undertaken by SAKSS at the country level. Box 1.1 provides the results of analysis for Senegal, showing the agricultural growth required to meet MDG1 by 2015, analysis of the potential sources of that growth within the agricultural sector, and the potential for increasing crop productivity. Box 1.2, on the other hand, uses the development domain concept to identify potential ARD strategies by agricultural commodity in Uganda. At the multicountry level, the tracking of expenditures in agriculture by a Regional Strategic Analysis and Knowledge Support System (ReSAKSS) is illustrated in Box 1.3. Key elements of the strategic analysis agenda of a ReSAKSS are helping to monitor changes in key factors that affect the outcomes of ARD strategies and encouraging information sharing, mutual learning, and cross-country comparisons.

The knowledge support system component

The strategic analysis is carried out within the context of a *knowledge support system,* which for SAKSS describes the network of individuals and institutions that serve the need for strategic information during the formulation and implementation of ARD strategies. The network includes a range of individuals (researchers, policymakers, and development practitioners) and organizations (government agencies, research institutes, universities, development organizations, and private-sector and civil-society groups). These individuals and organizations are linked through this network with the shared interest of seeking tangible solutions to the challenges facing ARD. Box 1.4 provides an example of a country SAKSS network, the types of members, and the various links among them.

Three core activities undertaken by a SAKSS network include collaborative strategic analysis, capacity strengthening, and dialogue. Through these activities relevant information from research findings and data analysis is compiled, synthesized, and packaged into evidence that enriches the dialogue on future ARD priorities in a timely, reliable fashion. The collaborative and participatory manner in which these activities are conducted is intended to help validate the relevant questions being asked by policymakers and civil society and the tools of analysis, data sources, and assumptions needed to address such questions; instill confidence in the evidence generated; and ultimately, enrich the capacity to generate and use analytical tools and evidence in the process of formulating and implement ARD strategies.

Applications of this strategy at the country and regional levels exist. For example, ReSAKSS has initially focused much of its attention on mobilizing networks of individuals and organizations associated with the CAADP agenda at the regional level. Through these networks and a number of workshop forums, ReSAKSS helps to fill critical knowledge gaps and bring together a stock of knowledge, expertise, and tools as countries begin to shape and align their agricultural strategies within the CAADP framework. A website brings together the information being collected to promote peer review across countries and updates on

Box 1.1: Strategic analysis: Prioritizing investments in agriculture

Example: The following analytical results use the example of Senegal to summarize the potential effect of productivity increases among key agricultural subsectors in terms of their contribution to overall sector growth and the achievement of poverty reduction targets. A regional economic multimarket model developed for West and Central Africa was used to estimate the individual crop-specific effects on overall sector growth among individual countries in the region. The analysis was part of a broader study that incorporated a range of analytical methods, such as development domain analysis (see example in Box 1.2), yield potential analysis within each domain, an economywide and regional economic analysis, and estimation of the effects of output growth on poverty. The overall objective of the study was to assess priorities and options for agricultural research and development in the West and Central Africa region.

Senegal will need to grow its agricultural sector faster, up to 6.6 percent, in order to meet the MDG1 target.

Simply closing current yield gaps will not be sufficient to ensure the meeting of MDG1 and broad-based poverty reduction.

Priorities: Cereals and livestock (stronger growth and poverty reduction); processed foods, vegetables, and oilseeds (some strong growth).

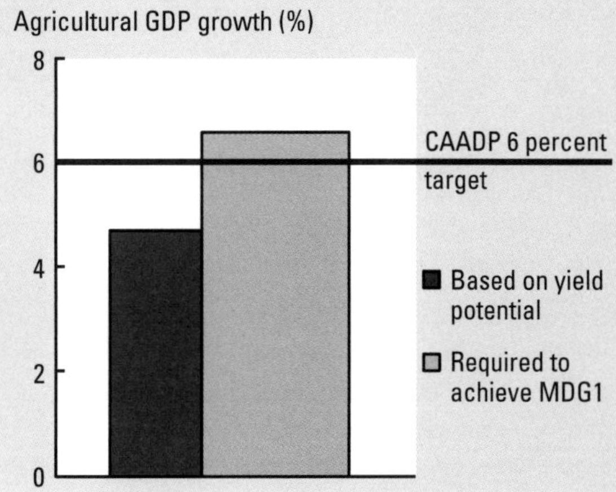

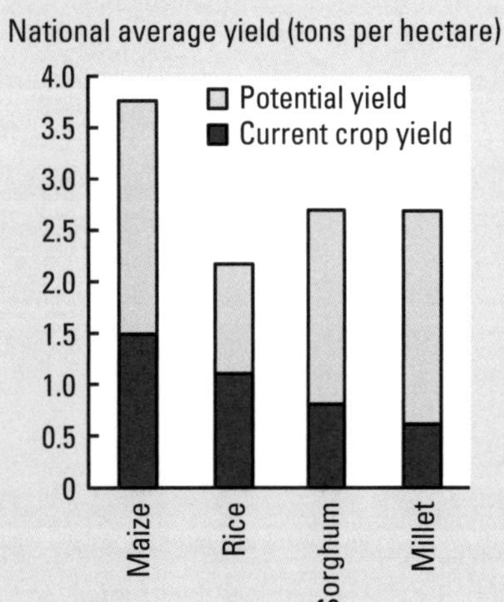

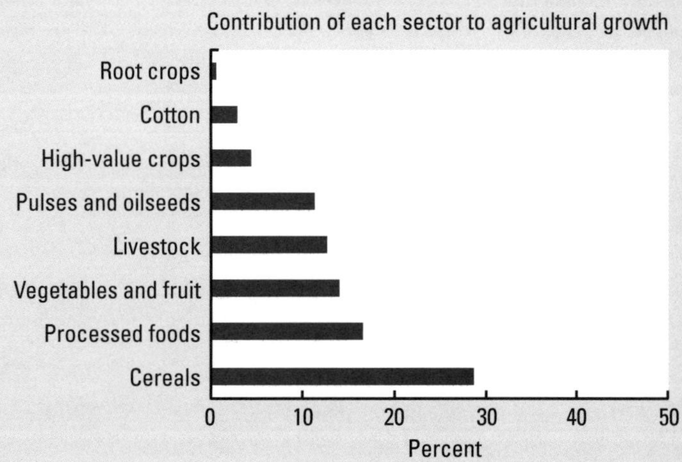

Source: Results summarized for Senegal are taken from Johnson et al. (2008).

Note: MDG means Millennium Development Goal; CAADP means Comprehensive Africa Agriculture Development Programme.

Box 1.2: Strategic analysis: Spatial analysis for targeting of investments by development domain in Uganda

The spatial analysis of development domains was undertaken as part of IFPRI's Strategic Criteria for Rural Investments in Productivity Project in Uganda, a precursor to the Uganda Strategic Analysis and Knowledge Support Systems. The agricultural development domains, as defined here, help to distinguish geographic areas where the intersection of population density, agricultural potential, and market access are distinctively similar, and therefore areas likely to have similar relevance for specific agricultural development strategies. The examples are for two extreme cases—high population density/high market access/high agriculture potential versus low population density/low market access/low agriculture potential—and corresponding potential agricultural development strategies based on local expert and literature reviews (see also Wood and Chamberlin 2003). For other resources on development domains, see www.harvestchoice.org.

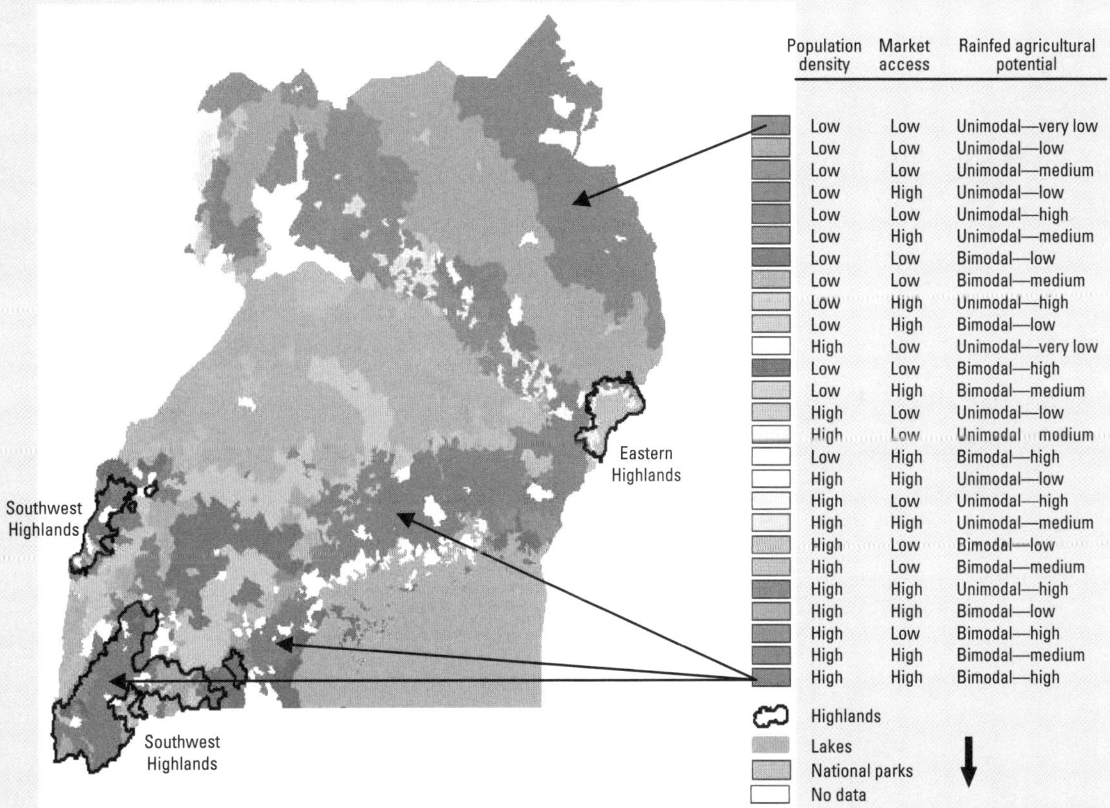

Potential agriculture development strategies (for example, for high/high/high and low/low/low)

Population density	Market access	Agricultural potential	Potential development strategy
High	High	High	Perishable cash crops
			Dairy, intensive livestock
			Nonperishable cash crops
			Rural nonfarm development
Low	Low	Low	Low-input cereals
			Limited livestock intensification
			Emigration

Source: Bolwig, Hazell, and Wood (2002).

Box 1.3: Strategic analysis: Tracking agricultural development goals

The example here shows how the ReSAKSS helps to track CAADP shared goals of allocating at least 10 percent of spending to agriculture and achieving an agricultural growth rate of at least 6 percent per year by 2008. These benchmark targets are sometimes referred to as part of the 2003 Maputo Declaration, in which African governments committed to these goals. The Declaration directed all African Union member countries to increase investment in the agriculture sector to at least 10 percent of their national budgets by 2008. The charts below review the progress to date against these benchmarks and at the Africa-wide level. Data have been collected from both in-country and international sources.

At the continental level, agricultural spending nearly doubled between 2000 and 2005.

In 2003, only 3.2 percent of countries allocated 10 percent or more of their budgets to agriculture.

This increased to 33.3 percent in 2006 before slightly falling to 25 percent in 2007.

Nine countries increased their allocations from less than 5 percent spending to 5–10 percent spending.

Level of agricultural spending as a share of total spending, 2002–07

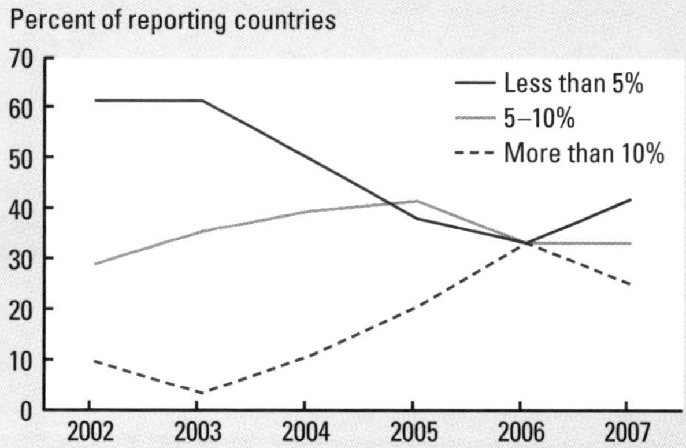

Average annual agricultural GDP and GDP growth rates in comparison to CAADP 6 percent target, Sub-Saharan Africa, 1990–2008 (%)

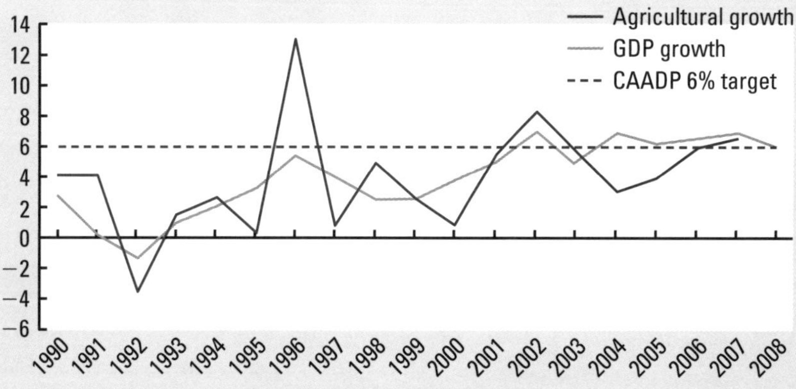

Sources: Fan, Omilola, and Lambert (2009); <www.resakss.org> (accessed July 28, 2009).

Example: Tracking Comprehensive Africa Agriculture Development Programme (CAADP) goals by Regional Strategic Analysis and Knowledge Support System (ReSAKSS).

Box 1.4: Country Strategic Analysis and Knowledge Support System (SAKSS) network

The SAKSS network is a set of key actors (individuals, organizations, and institutions) actively involved in providing strategic analysis, knowledge management, and capacity strengthening in the country to inform agricultural and rural development (ARD) strategies. Using the SAKSS operational and governance structure presented in Chapter 4 and illustrated in Figure 4.1, the different actors and their linkages in the network can be described in terms of how they operate to promote greater linkages between research evidence and decisionmaking for ARD.

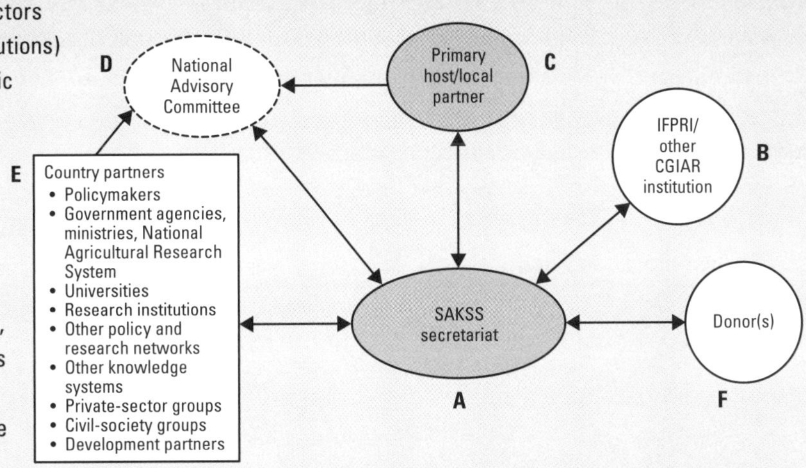

First, the SAKSS secretariat (A) is typically set up as a program administered by an international research organization such as IFPRI (B) but hosted by a local institution and/or collaborator (C) that is well regarded by policymakers (it may also be situated in the Ministry of Agriculture). Typically, the host institution is an active partner of the country SAKSS and thus serves on the National Advisory Committee, or NAC (D). The NAC is typically made up of collaborating local research institutions and individuals, policymakers in government ministries and among donors, and key private-sector and civil-society groups (E). The SAKSS network is composed of the NAC members, other national and international research collaborators who offer technical and training expertise, and other nonresearch groups including policymakers and members of the private sector and civil society, who ensure that the information generated and compiled is always relevant and timely.

Using the example of Ghana (see also Figure 5.2), these include:

A. The Ghana Strategy Support Program (GSSP), headed by a program leader (IFPRI senior researcher)

B. The International Food Policy Research Institute (IFPRI)

C. The Council for Scientific and Industrial Research (CSIR)

D. The NAC of GSSP, which is currently chaired by the minister of finance and economic planning

E. Other collaborating country partners, including:

- *National partners:* The Ministry of Food and Agriculture; the Office of the President; Ghana Statistical Services; the National Development Planning Commission; the Bank of Ghana; the Institute of Statistical, Social, and Economic Research of the University of Ghana; and the Private Enterprise Foundation. Not in the NAC are other partners that are periodically consulted such as national agricultural research institutes, universities, nongovernmental organizations (NGOs), and private-sector, civil-society groups.
- *International partners:* Bilateral (the U.K. Department for International Development, EU, and Canadian International Development Agency) and multilateral (the World Bank) donors, the Consultative Group on International Agricultural Research centers (for example, the International Water Management Institute), development practitioners (Chemonics and the International Center for Soil Fertility and Agricultural Development), and others periodically consulted, such as international NGOs, private-sector groups, data and information systems, universities, and research organizations

F. The primary donor, the U.S. Agency for International Development

Source: DSGD (2006).

Box 1.5: Knowledge support systems: Examples of the ReSAKSS website

The Regional Strategic Analysis and Knowledge Support System (ReSAKSS) website has been developed to provide easy access to information relevant to tracking progress of key benchmarks and development goals of the Comprehensive Africa Agriculture Development Programme (CAADP) of the African Union and the New Partnership for Africa's Development. In essence, it allows ReSAKSS to provide information on a country's track record in terms of agricultural and economic performance, CAADP benchmarks for agricultural spending, and the Millennium Development Goals for reducing poverty and hunger. It also provides access to new evidence from research intended to help guide future strategic directions for agriculture and rural development strategies. A blog site also promotes feedback.

a. Snapshot of a country page (Tanzania)

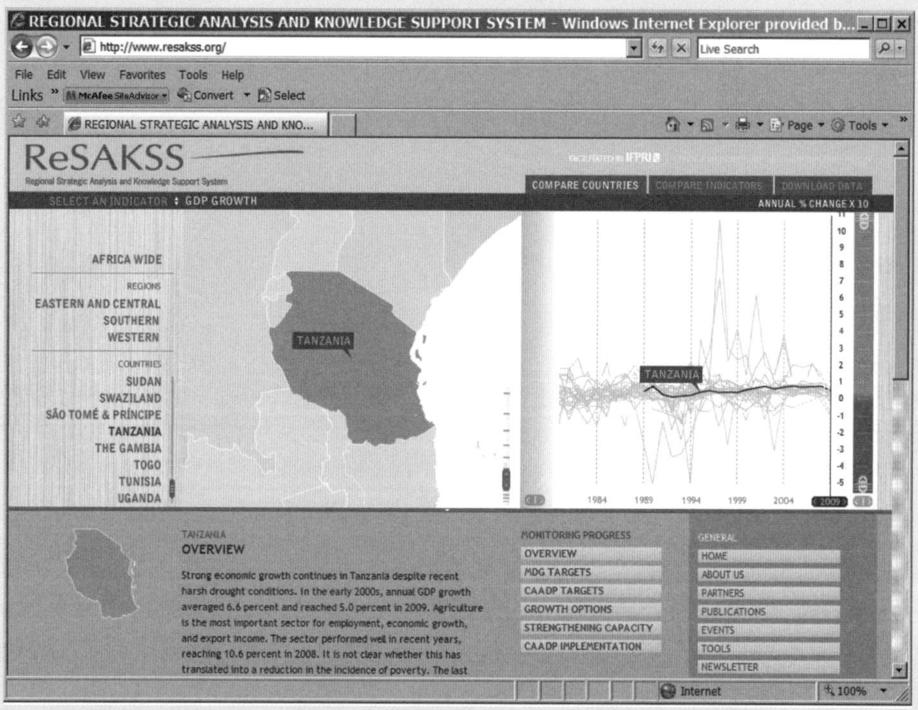

b. Snapshot of the ReSAKSS blog site (ReSAKSS e-newsletter)

Sources: Screen shots from <www.resakss.org> and <resakss.worldpress.com> (accessed July 28, 2009).

Box 1.6: Knowledge support systems: Examples of the Ghana Strategy Support Program website

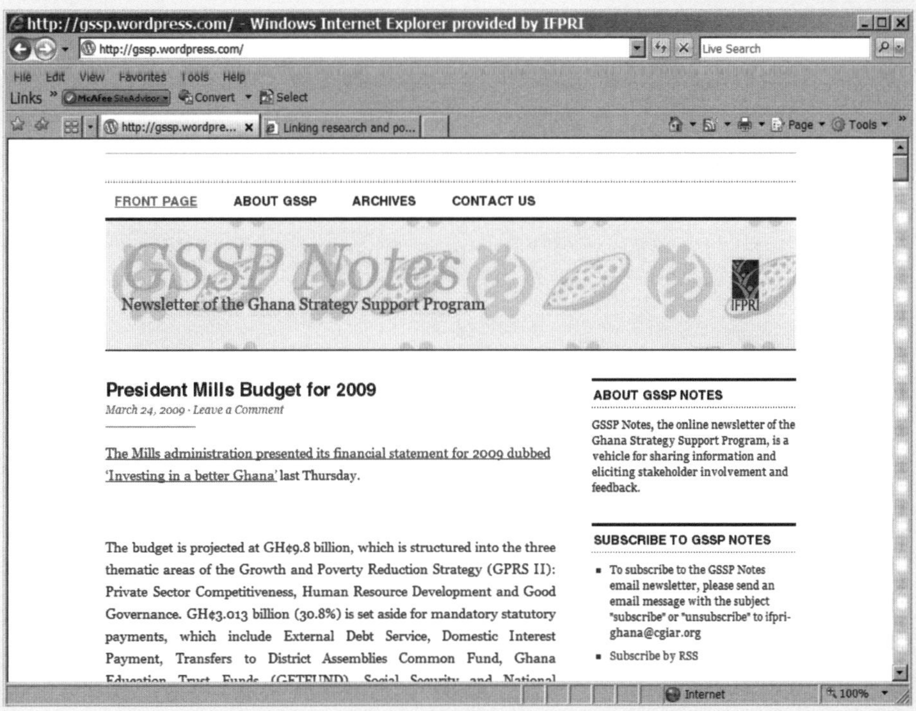

Sources: Screen shots from <www.ifpri.org/gssp> and <http://gssp.worldpress.com> (accessed July 28, 2009).

the progress of CAADP implementation (Box 1.5). The networks, in turn, are helping to strengthen local capacities to produce and apply credible evidence during policy dialogues about future ARD priorities. At the country level, information sharing is also organized via a website and a number of in-country events (Box 1.6 shows examples for Ghana).

Overall Goals

The principal goals of SAKSS are threefold:

- Bring quality and *strategic analysis* to bear on identifying key investments, institutional mechanisms, and policy options, as well as the implementation of selected options for ARD strategies. This includes harmonizing and generating standardized information for development planning and M&E to be available as *global public goods.*
- Build and strengthen *national and regional capacities* for policy analysis, monitoring, and evaluation while helping to bridge the research and policy divide for ARD strategies.
- Strengthen *dialogue* and links among institutions and individuals who supply and use data and information related to ARD strategies by establishing a network for information exchange and knowledge management.

Table 1.1 What SAKSS is and is not

Realm	What it is	What it is not
Overall	A collaborative network that acts as a mechanism by which relevant evidence is generated and used to inform agricultural and rural development (ARD) strategy formulation and implementation. It brings together two important concepts—"strategic analysis" and "knowledge support systems."	A silver bullet and a "one-size-fits-all" approach to informing the design and implementation of ARD strategies.
As a strategic analysis approach	A framework of analysis that helps identify policy and investment options for achieving high-end ARD goals (see Figure 2.1). The analysis uses a combination of tools, approaches, and synthesis in a flexible manner in order to consider diverse local circumstances and needs with respect to capacity for generating and using analytical evidence, extent of data availability, existing knowledge gaps, national goals, and timing of the strategy design and implementation process.	A single integrated model, analytical framework, or prepackaged and comprehensive toolbox for assessing investment and policy priorities for achieving high-end ARD goals.
As a knowledge support system	A network to serve the evidence needs of ARD strategy formulation and implementation (see Figure 3.1). This network includes individuals such as researchers, policymakers, and development practitioners and institutions such as government agencies, research institutes, development organizations, and private-sector and civil-society groups (see also Box 1.4). These individuals and institutions are linked through collaborative strategic analysis, capacity strengthening, and dialogue. Through this network, information, data analysis, and knowledge can be compiled, synthesized, and packaged into evidence that is supplied on a timely and reliable basis for use during strategy processes.	A top-down "development planning" and "monitoring and evaluation" system to serve the interests of national governments and donors while promoting policy research and analysis as an end in itself.
As a laboratory for research	An experimental and learning-by-doing environment for researchers interested in improving our understanding of how to bring credible evidence and analysis to bear during the process of designing and implementing ARD strategies, including the alternative approaches for doing so under different circumstances.	A single "how-to" manual on bringing research to bear on the development and implementation of all development strategies.

Source: Authors.

Conceptually, these goals are broad by design to allow sufficient flexibility as to how a SAKSS network can evolve and adapt to local conditions and needs while still maintaining a public goods nature. Given the rather abstract nature of SAKSS, however, it is important to consider its boundaries by summarizing what it is and is not, which we have done in Table 1.1. Finally, how the hefty goals of SAKSS have been translated and adopted on the ground, either as part of country strategy support programs of the International Food Policy Research Institute or as part of the multicountry ReSAKSS effort, has differed depending on local demand, existing institutional linkages and capacities for data collection and analysis, and the level of interest and funding within each country.

The Purpose of This Technical Guide

The purpose of this technical guide is primarily to define the concept of and present practical guidelines for "strategic analysis" approaches and the setup of "knowledge support systems," or SAKSS, based on both theory and lessons being generated from existing applications. The need for such a technical guide has come about due to a growing demand for the kinds of applications offered by SAKSS as many more African countries express their wishes to establish similar knowledge support systems. This demand is occurring at a rapid pace because governments are challenged with remaining committed to shared development goals (for example, achieving the MDG and CAADP goals). The endorsement of the SAKSS concept by NEPAD, African governments, the donor Group of Eight, and other development partners has also fueled the demand. The establishment of the ReSAKSS is a response to this growing demand to facilitate and guide the establishment of country SAKSS as an integral part of the CAADP agenda.

The technical guide is therefore intended to serve

- researchers and policy analysts, research institutions and think tanks, and government technical agencies, which are helping to set up country SAKSSs;
- African government and development partner policymakers who wish to use SAKSSs for their evidence needs;
- other policy analysts and researchers who simply wish to participate in such a network; and
- farmer and trader organizations, practitioners, policy and development advocacy groups, NGOs, media organizations, and private-sector and civil-society groups—as key stakeholders and beneficiaries—who may wish to know more about SAKSS.

This technical guide has been organized into three main parts to allow readers to skip to the appropriate part depending on their expertise, interests, and backgrounds. The first part (Part I) is about theory and methods. It provides a background for the definition of the SAKSS concept and the approaches used, beginning with strategic analysis (Chapter 2) and followed by the knowledge support system framework (Chapter 3). The second part (Part II) is far more practical in nature. It provides a guide on how to set up a country SAKSS (Chapter 4), followed by a review of a number of existing applications (Chapter 5). A conclusion highlights some of the lesson to date while posing some challenging questions for future learning and applications. The final part (Part III) offers appendixes of tools and resources. Although these are not exhaustive, they illustrate the types of tools useful for strategic analysis, as well as providing some practical resources: the terms of reference of a SAKSS program coordinator, how to develop a SAKSS communication strategy, and a glossary of related terms.

Part I: The SAKSS Concept and Approach

Chapter 2: The "Strategic Analysis" Concept and Approach

In the context of SAKSS, *strategic analysis* describes a series of analyses that are intended to help narrow investment and policy options for ARD in order to achieve growth and poverty goals.[5] For most developing countries, goals might include the MDG of halving poverty by 2015 (MDG1), a certain increase in average incomes per year, or, in line with the CAADP target, a 6 percent agricultural growth rate. The analysis is considered strategic as long as it contributes to the narrowing down of investment options that will help lead to the achievement of these high-end development targets. That means it must weigh in the costs and benefits of undertaking one strategy over another. Should government focus on promoting a rapid expansion in food staples production? It will definitely benefit consumers, but producers could be devastated if prices drop too fast following a bumper harvest. Or governments may wish to introduce a policy that is designed to target the poor and yet undercuts private-sector participation and long-term sustainability, as in the case of output procurement and the provision of modern inputs.

The strategic analysis of a SAKSS consists of answering a series of logically sequenced questions that help guide the analysis and lead to the identification and implementation of key investment and policy options for achieving national goals and targets. The sequence and types of analysis are not fixed, however. Different local contexts may require a different set of analyses. Because SAKSS has been primarily developed for African countries with a large agricultural sector, the analysis is focused on identifying options for ARD as a source of economic growth and poverty reduction.

In this chapter we review typical strategic-level questions that arise in considering future priorities for ARD. We summarize these in Figure 2.1 for quick reference. Answers to these questions can contribute positively during the process of formulating and implementing an ARD strategy. There are various economic tools and methodologies that can be used to analyze some of these questions further, with varying degrees of sophistication. Which tools and approaches are used will depend not only on the question being asked but on many other important considerations as well, such as the availability of data and expertise, the time available to undertake the analysis, cost, access to analytical tools and economic models, and underlying assumptions and limitations. Depending on the level of sophistication required in the analysis, international experts are often needed to undertake the analysis, which can add to the total costs of the analysis very quickly. Appendix A summarizes these tools in more detail. Where there are already existing studies and sufficient evidence to draw from, new analysis may not be necessary.

There are also a number of useful sources that provide a range of economic tools and analytical methods to inform and guide development strategy. For example, associated with supporting the development and implementation of poverty reduction strategies, a World Bank poverty and social impact assessment source book (World Bank 2003) reviews a wide range of tools and their applications. The source book provides guidance as to which tool is most appropriate under what conditions with respect to a range of factors such as data availability, cost, and time to complete the analysis. Bourguignon and Pereira da Silva (2003) also provide an extensive review of the range of available evaluation techniques and tools for assessing data and the impact of policies on poverty reduction, from estimating geographically disaggregated welfare levels and changes to predicting the effects of aggregate growth on poverty. Finally, Bandara (1991) reviews the application and use of the more sophisticated computable general equilibrium models, especially in the context of policy analysis in developing countries.

The chapter is organized and sequenced around four goal-level questions for ARD (as illustrated in Figure 2.1), each with a corresponding series of follow-up questions that help provide key evidence to inform the *local political process* of designing and implementing an ARD strategy:[6]

1. What are the economywide options and trade-offs for reaching high-end national development goals through an ARD?

2. How can ARD strategies be targeted to address the diversity of opportunities and challenges within a country?

[5] Chapter 2 is partially based on an earlier paper by Johnson and Resnick (2004).

[6] The emphasis on political process is purposeful. Local stakeholders and policymakers must be consulted throughout the process of generating and disseminating results of analysis, not only in order to validate the assumptions and data of the analysis but also to ensure that the questions being addressed by the analysis are timely, salient, and relevant. This process is described further when discussing the "knowledge support system" component of a SAKSS in the next chapter.

Figure 2.1. Strategic analysis to inform agricultural and rural development (ARD) strategy

Setting the context: **What are the economywide options for reaching high-end national development goals through an ARD strategy, and what are the trade-offs involved?**

- What is the current state of affairs or the initial situation with respect to development? Is the country on track to achieve its national growth and poverty goals?
- What are the contributions of different sectors to growth and poverty reduction? What is the role of agriculture in the economy?
- Where agriculture is a key sector, what level of effort and performance is required to meet the overall growth and poverty reduction targets?
- What are the key sources of growth among agricultural subsectors, and which of them are more likely to be pro-poor?
- Are there sufficient demand and market opportunities to absorb any rapid increases in supply within selected subsectors? What are the key constraints on commercialization?
- Are there opportunities for leveraging regional growth dynamics through greater multicountry cooperation and economic integration?

Spatial considerations: **How can ARD strategies be targeted to address the diversity of opportunities and challenges within a country?**

- What is the distribution of income, and distribution and extent of poverty and malnutrition, across different locations in the country?
- What kinds of opportunities and challenges affect rural economic livelihoods in different parts of the country?
- For agriculture, what are the key development domains, as defined by agricultural potential, market access, and population density?
- Which development domains offer the greatest potential for high investment impact among the key subsectors and economic activities identified as key sources of growth in the economywide analysis above?
- What kinds of interventions (for example, infrastructure, R&D and extension, institutional) and policy reforms are needed to spur productivity and income growth among select domains and/or particular commodity subsectors?
- Among the poorest of the domains, with the most limited resource assets and livelihood options, what are the alternatives for poverty reduction and food security?

Weighing investment priorities: **How should resources be mobilized and allocated across the different economic sectors and geographic regions?**

- What have been the trends in government expenditures by sector, and what have been the reasons for these changes?
- How has public investment been financed, and how has the burden of financing investment been distributed in society?
- What have been the economic rates of return to various types of government expenditures, including their impact on growth and poverty reduction?
- What level of public spending is required to achieve targeted goals for agriculture and overall economic growth?

Monitoring and evaluation: **What have been the impacts of ARD strategies, and what lessons do they offer?**

- Are the allocation and level of inputs (for example, spending, investments, policy interventions) of the ARD strategy (or project) on target?
- How can the impact of these inputs be traced to improvements in outputs (for example, productivity and the viability of production systems, food processors, agroindustries, markets, and trade)?
- Have these improvements affected outcomes (for example, incomes and the poverty status of target populations)?
- What factors have shaped (positively and negatively) the impacts achieved to date? What needs to be altered? What was the distribution of these impacts (for example, did they affect smallholders, equity, or gender or have other spillover impacts)?
- Can key ingredients of success or failure be identified based on past experience and lessons learned? What have been the roles of the public versus the private sector?

Source: Authors.

3. How should resources be mobilized and allocated for ARD across the different economic sectors and geographic regions?
4. What have been the lessons and effects of ARD strategies on outcomes and impact?

Because learning from the past impact of ARD interventions is crucial to improve the design of future strategies (thus completing a full circle), a significant portion of this chapter has been devoted to the last question. The discussion under each of these four questions is organized around providing answers to four additional questions: Why is this question important? What are some corresponding questions for strategy? What is needed to answer these questions? Are there applications to draw from?

WHAT ARE THE ECONOMYWIDE OPTIONS AND TRADE-OFFS FOR REACHING HIGH-END NATIONAL DEVELOPMENT GOALS THROUGH AN ARD?

Why is this question important?

Most developing countries have set targets for achieving high-end development goals such as economic growth, poverty reduction, and food security. From the outset, therefore, it is useful to first establish a country's current situation and whether its trajectory will lead to the achievement of its goals. This should be done within the context of the country's overall economy in order to highlight a broad set of strategic options and trade-offs, for example, whether simply promoting faster growth is more important than considering poverty and food security or even environmental degradation. This context is needed because policies at the macro level, such as those for trade and market liberalization, can have a profound impact on growth and an even greater effect on agriculture, the rural economy, and poverty (Dorward et al. 2004). At the same time, policies that directly affect rural areas and agriculture can have an impact on the overall economy and in turn have feedback effects on the rural sector. By examining many of these policy options within the context of the broader economy, key relationships and welfare implications can be assessed in ways that lessen any potential adverse impacts on the poor.

The economywide perspective, therefore, permits higher-level strategic questions to be posed when shaping an agriculture or rural development strategy within the context of overall national development goals and thus provides the greatest strategic leverage to priority setting (Byerlee 2000). The potential contribution of agriculture to economywide growth and national development priorities such as reducing poverty can then be explored. This exploration could inform national debates over broader development strategies. One such broader approach to development is the poverty reduction strategy process introduced by the World Bank and IMF in 1999 in order to make aid more effective by requiring recipient governments to direct resources toward poverty-reducing public-sector programs (World Bank 2004b). Within this normative mode of analysis, questions regarding the long-term distributional consequences of alternative investment and policy choices for meeting these targets can also be explored. Specific to rural-sector strategies, sectorwide investment options should be examined more closely, especially with regard to how they affect the incentives for rural agricultural production and commercialization.

What are some corresponding questions for strategy?

Exploring a country's economywide options and trade-offs for reaching high-end development goals introduces a series of strategic questions that require answers, such as these:

- What is the current state of affairs or the initial condition with respect to development? Is the country on track to achieve its national growth and poverty goals?
- What are the contributions of different sectors to growth and poverty reduction? What is the role of agriculture in the economy?
- Where agriculture is a key sector, what level of effort and performance is required to meet the overall growth and poverty reduction targets?
- What are the key sources of growth within agricultural subsectors, and which ones are more likely to be pro-poor?
- Are there sufficient demand and market opportunities to absorb any rapid increases in supply within selected subsectors? What are the key constraints and/or bottlenecks to commercialization?
- Are there opportunities for leveraging regional growth dynamics through greater multicountry cooperation and economic integration?

What is needed to answer these questions?

In addition to a situational analysis of a country's current development status and progress toward achieving its national development goals, an economywide simulation can help us to explore the strategic questions just raised. The computable general equilibrium (CGE) model is particularly well suited to this purpose (see first section of Appendix A). CGE models help analyze the effects of policy shifts and alternative sector growth scenarios on overall economic growth and poverty reduction. They have the advantage of capturing both direct and indirect effects of policy changes on poverty and income distribution given a country's overall economic structure. The effects are channeled through changes in employment, wages, and relative prices while considering forward and backward linkages in the economy. From this information policymakers can weigh the costs and benefits associated with focusing attention on stimulating growth in different sectors and subsectors. For example, faster growth may be possible in higher-value export sectors; however, the poverty-reducing effect of such growth would be limited. The cost to society would be the reduced effect on poverty.

Data sources for economywide analyses are typically found in countries' ministries of trade and finance and customs, as well as national reserve banks, which are principal sources of trade and macroeconomic information. Much of this information is used in creating social accounting matrixes (SAMs), which provide the basic information for macroeconomic and CGE models. Unfortunately, there are few SAMs that are maintained or updated frequently in Africa. This is due not only to a lack of adequate and relevant data but also to a lack of capacity to generate them. Their use is even more limited because of poor capacity but also because of general skepticism about its value. Usually donors are more apt to pay for them, including the application of CGE models.

However, it is not always necessary to rely on a full SAM and/or CGE model. Where data are limited and a CGE model is not available, a simpler economywide multimarket model can be used in its place (see second section of Appendix A). Although less sophisticated, this model is actually better suited to answer questions specific to ARD. It can treat the agricultural sector in more detail by incorporating useful market and trade linkages across various commodities and locations and by combining a system of demand and supply equations that allow for interactions across commodities. To partially maintain an economywide perspective, a nonagricultural sector is included in the model to capture potential agricultural linkages with this sector.

Are there existing applications to draw from?

A number of examples that apply this level of strategic and economywide analysis involve the work that has been undertaken by researchers at the International Food Policy Research Institute (IFPRI) in a number of countries in Africa (for example, Ethiopia, Ghana, Malawi, Mozambique, Rwanda, and Zambia). From these studies it was evident that most countries would not meet MDG1, with the exception of Ghana, Mozambique, and Uganda (Box 2.1). For Mozambique, for example, Thurlow (2008) has found that although the current growth rates in agriculture will not be sufficient to meet the national goal of halving poverty by 2015, the potential is there at higher rates of growth. Model results further show that the additional growth will need to be driven by maize and other cereal crops if it is to achieve this goal, because growth in the cereals sector has a larger impact on poverty reduction than does similar growth in export-oriented crops (see Box 2.2). This impact occurs because yield improvements in foodcrops benefit households not only directly, by increasing incomes from agricultural production, but also by allowing farmers to diversify into higher-value crops. Cereals are already an important sector in Mozambique and have strong growth linkages to nonagriculture, which stimulates broader economywide growth and poverty reduction.

In the example of Uganda, the analysis shows that although the staples crop sector has greater poverty-reducing effects, the export crops sector still has a significant effect on overall growth. So if the government is more concerned about poverty reduction, focusing attention on cereals and root crops will have a greater impact. On the other hand, if policymakers are more concerned about growth, both the export and root crops sectors will contribute more to growth. The cost in focusing on export crops alone would be a reduction in the rate of poverty decline in the country (see Box 2.2). For Ethiopia, it is clear that although agriculture is important for overall growth and poverty reduction, the effects are even greater if there is simultaneous growth in nonagriculture as well. Within agriculture, food staples and livestock would offer the greatest impact on agriculture growth and poverty reduction (see Box 2.1).

HOW CAN ARD STRATEGIES BE TARGETED TO ADDRESS THE DIVERSITY OF OPPORTUNITIES AND CHALLENGES WITHIN A COUNTRY?

Why is this question important?

Many of the challenges and opportunities that national development strategies must negotiate are geographic in nature and affect different communities on the ground. For example, economic opportunities can vary widely by location depending on other important factors such as the natural resource base (for example, agriculture potential), population density, and access to markets and rural services (for example, education, health, and agriculture extension). Development options would be quite different for more remote and food-insecure areas versus those located in close proximity to large market centers (Box 2.3). Thus any interventions should be specifically

Box 2.1: The role of agriculture in growth and poverty reduction among selected countries

As part of the Regional Strategic Analysis and Knowledge Support System analytical support to the Comprehensive Africa Agriculture Development Programme (CAADP), a number of researchers at the International Food Policy Research Institute were asked to determine whether a 6 percent growth rate for agriculture was suffcient to generate overall growth and the achievment of the Millennium Development Goal of halving poverty by 2015 (MDG1). The results from these four country case studies are based on economywide computable general equilibrium models that analyzed the extent to which agriculture can drive overall growth and the extent to which such growth ultimately impacts poverty. The results show that agricultural growth has a very strong poverty-reducing effect. However, most countries will need to grow much faster than CAADP's 6 percent growth target in order to achieve MDG1. The exceptions are Mozambique and Uganda.

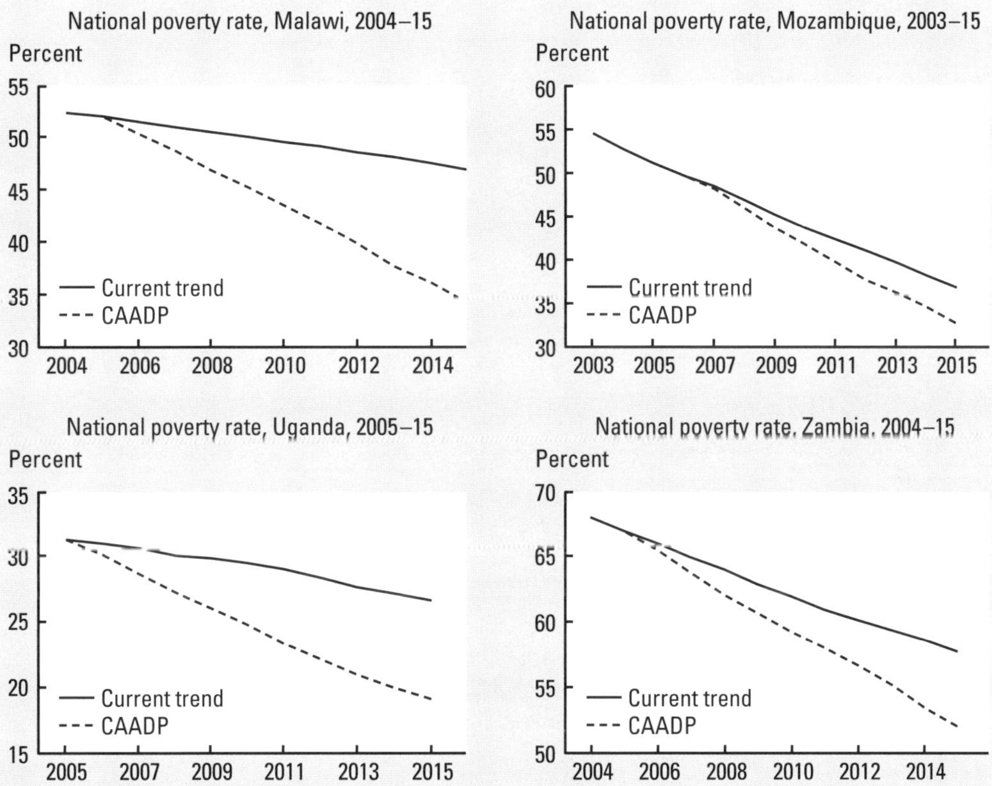

Sources: Benin et al. (2008a) for Malawi; Thurlow (2008) for Mozambique; Thurlow et al. (2008a) for Uganda; and Thurlow et al. (2008b) for Zambia.

targeted at the unique characteristics of an area and the severity of any problems there; quantifying the extent and distribution of poverty and malnutrition across geographic areas and population groups is an important first step (Babu and Pinstrup-Andersen 1994).

With the increasing availability of spatially disaggregated data and tools to understand those data, it has become increasingly possible to map indicators of biophysical and socioeconomic factors showing local comparative advantages for different agricultural and rural development options (see

Box 2.2: Analysis of sources of agriculture growth necessary to achieve the MDGs and the CAADP target

The results from these two case studies are based on an economywide analysis intended to answer these questions: Can the country achieve the Comprehensive Africa Agriculture Development Programme (CAADP) target for growth and Millennium Development Goal 1 (MDG1) of halving poverty by 2015? Which subsectors in agriculture contribute the most to growth and to poverty reduction?

(a) Mozambique

- On track to meet MDG1, *but* poverty is high, especially in the north, and agriculture's postwar "bounce-back" is slowing down
- Very low yields for most crops but high agroecological potential; large biofuel potential (that is, land and water resources and climate) *and* pro-poor if smallholder based (for example, outgrower *jatropha*)

- Strong export potential (that is, coastal access) but poor infrastructure

Priority crops: Maize and biofuels (strong growth and poverty effects)

RANKING SUBSECTORS AND CROPS

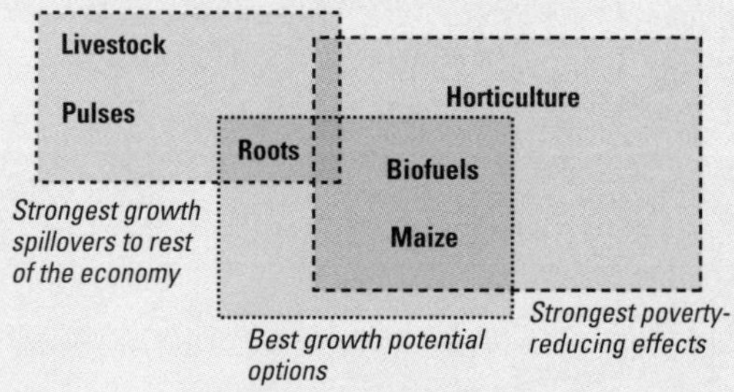

REDUCTION OF NATIONAL POVERTY RATE UNDER CAADP, 2003–15

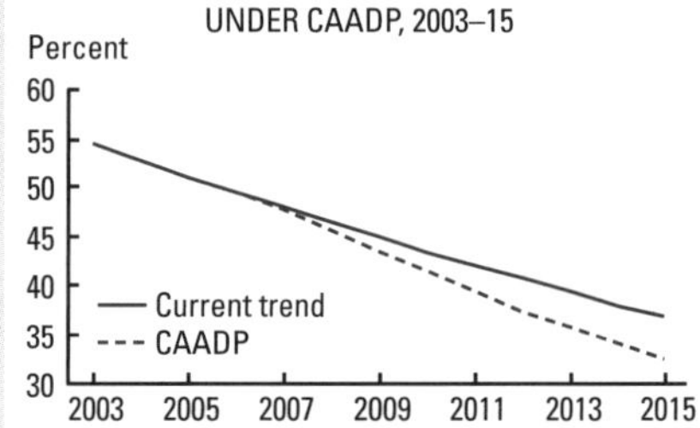

(b) Uganda

- Uganda will come close to missing MDG1, and poverty in the north will remain high
- Diverse agricultural sector, but large yield gaps persist for major crops (especially foodcrops)
- Closing yield gap ensures meeting MDG1 and creates broad-based poverty reduction

Priority crops: Root crops and new export crops (strong growth, poverty, and spillover effects)

RANKING SUBSECTORS AND CROPS

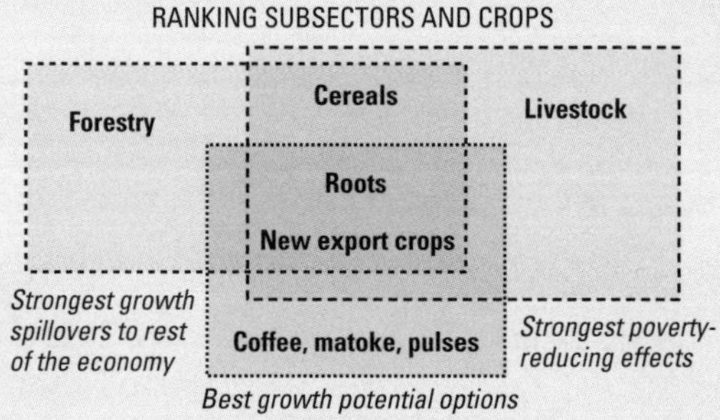

NATIONAL AVERAGE YIELD

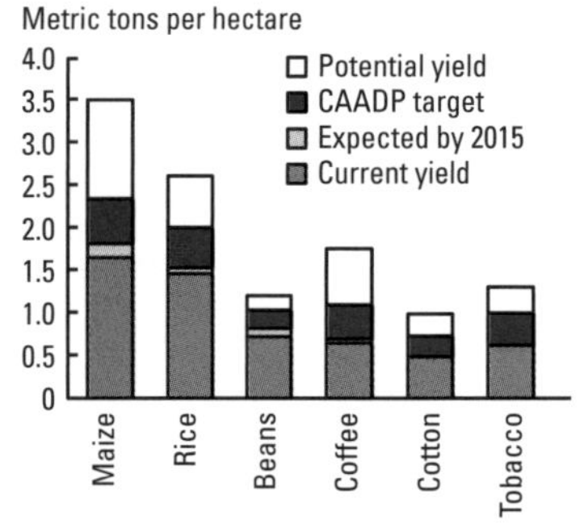

REDUCTION OF NATIONAL POVERTY RATE UNDER CAADP, 2005–15

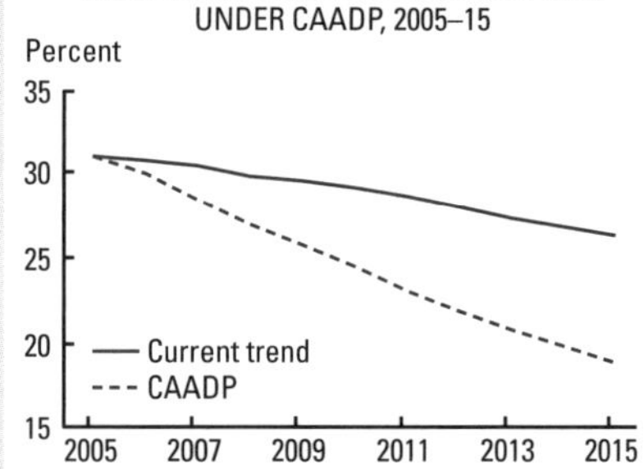

Sources: Thurlow (2008) for Mozambique; Thurlow et al. (2008a) for Uganda. See Box 2.1 for background.

Box 2.3: Market accessibility in Uganda

Market accessibility is a key concept in the analysis of actual and potential development patterns and is strongly linked with the kinds of livelihood strategies that people are able to pursue in different places. Mapping the areas of high versus low access in a country can yield a very powerful characterization of heterogeneous development options and is often incorporated into the definition of recommended domains for different policies. Such characterizations are often measured in terms of physical access to targets that are proxies for market opportunities (for example, estimated travel time to towns of more than 100,000 persons) and are designed to be relevant to generalized agricultural inputs and outputs.

However, there are several ways in which to take a more nuanced view of access to markets, which have implications for how it can be represented, both conceptually and in space—that is, in cartographic models that enable quantitative analysis.

In Uganda, for example, one can distinguish among

- local markets where low-value foodcrops are bought and sold (a);
- regional markets of nonperishable foodcrops accessed via the border with Rwanda and Congo and including perishables for Kenya and Tanzania (b);
- regional and international markets accessed via the Kigali–Kampala–Nairobi–Mombasa rail corridor for nonperishable exports (c); and
- "international fresh markets" for horticulture accessed via the airport at Entebbe (not included).

Considering the different commodities and the qualities or value added required by these markets, cartographic representations of physical access can further "spatialize" the relevant marketing chains, that is, mapping access to the sequential nodes of the entire production and marketing process. The result is a more complex but more directed set of indicators for policy analysis.

(a)

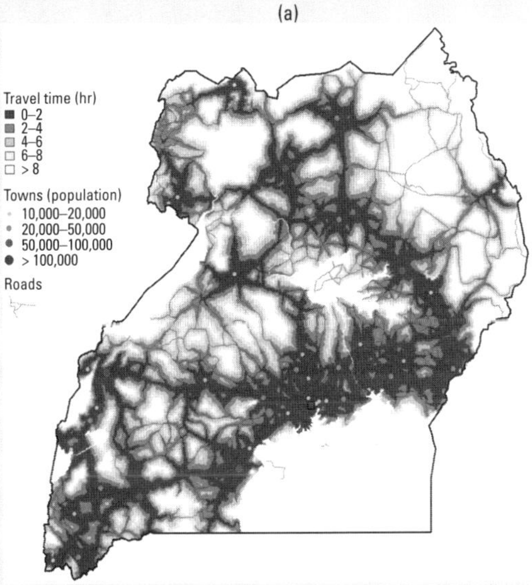

(b)

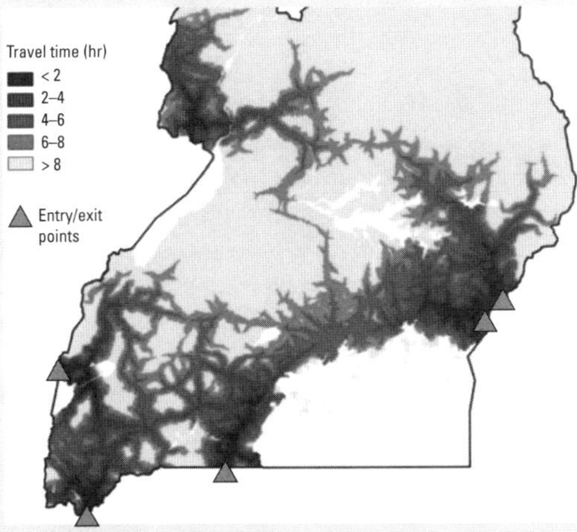

(c)

Source: Johnson and Resnick (2004, Box 4).

the example of development domains in Box 1.2). Agroclimatic factors, access to markets, and population density are some of the more important dimensions for assessing agricultural development potential (Pender et al. 2001). By viewing how these conditions correlate and overlap spatially with local welfare measures, assumptions can be made about how different development investments will impact the poor and what environmental costs there may be for changing agricultural land uses. Taken together, these conditions provide an enhanced picture of the costs and benefits of different investments, allowing better targeting toward the goals of sustainable growth, poverty reduction, and environmental sustainability (Wood et al. 1999; Wood and Chamberlin 2003).

To enable such decision support tools, there is a need for the compilation of relevant georeferenced and harmonized datasets with national coverage. In many countries, relevant data exist but are not often brought together for policy analysis. Furthermore, there is a need for analytical capacity (and knowledge of local conditions) to define geographic domains (or build other analytical frames) that reflect real constraints and opportunities for local development options.

Cross-sectoral perspectives should also be considered when analyzing spatially disaggregated data. For example, environmental factors may also affect agricultural productivity in various regions of a country. The analysis could look at the potential externalities of increased productivity related to the surrounding natural resources and assist in identifying "hot spots," that is, locations where the goals of economic development and natural resource conservation may be in competition with each other.

What are some corresponding questions for strategy?

Taking on a spatial perspective helps in targeting interventions. The typical questions to answer include these:

- What are the distribution and extent of income, poverty, and malnutrition across different locations in the country?
- What kinds of opportunities and challenges affect rural economic livelihoods in different parts of the country?
- For agriculture, what are the key development domains based on agricultural potential, market access, and population density?
- Which development domains offer the greatest potential for high investment impact among the key subsectors and economic activities identified as key sources of growth in the economywide analysis discussed earlier?
- What kinds of interventions (for example, infrastructure, R&D and extension, institutional) and policy reforms are needed to spur productivity and income growth among select domains and/or particular commodity subsectors?
- Among the poorest of the domains, in terms of limited resource assets and livelihood options, what are the alternatives for poverty reduction and food security?

What is needed to answer these questions?

Exploring answers to these questions requires sufficient access to spatially oriented data, including information on agroclimatic conditions, land use, production, urban and market centers, infrastructure, household consumption, and welfare. This information is often insufficient in developing countries, Africa in particular. However, with the increasing sophistication of computer technologies and satellite imagery, filling in some of the gaps is increasingly possible. For example, You et al. (2007) recently used a cross-entropy approach to make plausible allocations of crop production by small square grids (or pixels) based on available statistics for larger subnational units and satellite imagery.

Useful tools that help to map out geographically referenced data include geographic information systems (GIS) and remote sensing (see fourth section of Appendix A for more details). What is often missing, however, is a more integrated role for spatial analysis in targeting public-sector interventions. In addition to highlighting micro-level comparative advantage and thus livelihood options, the information can be integrated with economic behavioral and normative models to identify targeted investment and policy alternatives.

Data limitations will most likely define the type and degree of disaggregation possible, especially when dealing with socioeconomic data, which are often not georeferenced. It is therefore important to consider undertaking spatial analysis whenever the questions for strategy involve more micro-level targeting if the costs for doing so are not exorbitant. Such analysis must also complement other sectorwide, thematic (for example, physical infrastructure and policy environment), or economywide analysis to maintain a focus on ultimately achieving high-end national development goals.

Finally, for spatial analysis to prove useful in guiding strategic policy decisions, there are a few important considerations (Wood and Chamberlin 2003):

1. There must be an effort to assemble spatial information from a variety of sources.
2. Cross-sectoral expertise must be brought to bear on the development of meaningful spatial analysis and information tools.
3. The means of disseminating spatially based information to nonspecialist users must be identified.

4. Spatial information must find its way into existing government networks of information flow and mechanisms for inclusion in the strategic planning process. Georeferenced data are still limited in many countries, although this may be changing; increasingly, data collectors are beginning to use innovative georeferencing tools as they undertake surveys.

Are there applications to draw from?

A number of examples illustrate the usefulness and application of spatial analysis for targeting investments in agriculture. At the country level, useful examples include the work undertaken by Pender et al. (2001) and Bolwig, Hazell, and Wood (2002) to map out development domains in Uganda (see also Box 1.2). The more recent work by Chamberlin, Pender, and Yu (2006) built on this concept. At the regional level, the study by Omamo et al. (2006) is especially noteworthy. Spatial analysis tools were used to map out key development domains for agricultural production, market access, and population density across political boundaries in eastern and central Africa. This mapping was subsequently complemented by various economic analyses to assess future agricultural growth options and research priorities in the region. More recently, another good example was a study by Johnson et al. (2008) that used spatially generated yield potential estimates by development domain in West Africa to determine the impact on sector growth and poverty reduction of closing current yield gaps using existing technologies and know-how.

HOW SHOULD RESOURCES BE MOBILIZED AND ALLOCATED ACROSS THE DIFFERENT ECONOMIC SECTORS AND GEOGRAPHIC REGIONS?

Why is this question important?

An essential component of a development strategy is its plan for prioritizing investments and mobilizing resources. A strategy grounded in a country-specific context must be based on a thorough assessment of the public investment situation and the potential to contribute to progress toward development goals. Public investments can be thematic (for example, roads, marketing institutions), sectorwide (for example, research and extension, irrigation), and subsector specific (for example, commodity-based research).

All these investments affect rural poverty through many channels. For example, public investment in agricultural research, rural education and health, and infrastructure increases farmers' income directly by increasing agricultural productivity and lowering the transaction costs of both inputs and outputs, which in turn reduces rural poverty. Indirect impacts come from higher agricultural wages and improved nonfarm employment opportunities induced by growth in agricultural productivity and increases in market opportunities. Growth in agricultural output from rural investment often yields lower food prices, again helping the poor indirectly because they are often net buyers of foodcrops. Redistribution of land caused by higher agricultural growth also has important impacts on rural poverty. In addition to their productivity impact, public investments in rural education, health, and infrastructure directly promote rural wages, nonfarm employment, and migration, thereby reducing rural poverty. For example, improved infrastructure access will help farmers set up small nonfarm businesses in rural areas, such as food processing and marketing enterprises, electronic repairs shops, transportation and trade businesses, and restaurant services. A key underlying assumption is that public and private investments are complements (Anderson, de Renzio, and Levy 2006), so an increase in public goods and accumulation of capital stock increases the productivity of all factors in agricultural production, which in turn leads to higher farm wages and incomes and greater poverty reduction.

Investments in the rural sector not only contribute to growth, more employment opportunities, and higher wages in rural areas but also enhance the development of the national economy by providing labor, human and physical capital, less expensive food, and markets for urban industrial and service development. This type of growth in the national economy can then help reduce poverty in both rural and urban sectors. Understanding these different effects provides useful policy insights to improve the effectiveness of government poverty reduction strategies. In particular, it provides information on how public investment can be used to strengthen links among poverty reduction channels to increase efficiency in targeting public resources to poverty reduction. More efficient targeting has become increasingly important in an era of macroeconomic reforms, in which governments are under pressure to reduce budgets. For examples of tools and approaches to measure the impact of investments, see the fifth section of Appendix A and Benin, Johnson, and Omilola (2010).

What are some corresponding questions for strategy?

Answering the question of how resources should be mobilized and allocated across the different economic sectors and geographic regions essentially answers a range of high-end questions that inform the design and evaluation of a development strategy, such as these:

- What have been the trends of government expenditures by sector, and what have been the reasons for their changes?

- How has public investment been financed, and how has the burden of financing investment policy been distributed in society?
- What have been the economic rates of return to various types of government expenditures, including their impact on growth and poverty reduction?
- What level of effort in public spending is required to achieve targeted goals for agriculture and overall economic growth?

Analyzing these series of questions not only helps us identify the kinds of public-sector investments that offer the highest economic rate of return but also helps us assess the extent to which past investments have affected overall development goals, a topic we will cover later under the fourth and last goal-level question for ARD.

What is needed to answer these questions?

Answering the types of questions addressed by this analysis requires sufficient subnational data on the level and distribution of public-sector expenditures and investments over time. A public expenditure review is especially useful as a first step in compiling the required data. It will also help determine the extent to which actual resource allocations are consistent with a country's strategy and goals. Combining this information with other data, such as household survey data on consumption, production, and welfare measures, will allow for more sophisticated analyses. For example, using econometric tools, more detailed analysis can be carried out, drawing on the cross-sectional variation of the data, to measure differences in outcome variables such as growth and poverty and attribute them to the accumulated stock of past investments and other socioeconomic variables. Where time-series data are also available for the same cross-sectional data, the dynamics and lagged effects of public investments can also be analyzed. When these effects are combined with independent estimates of the unit costs of different investments, cost–benefit ratios can be calculated.

Results from the econometric analysis can be further translated into unitless elasticity estimates. An elasticity estimate measures the sensitivity or marginal effect of a 1 percent change in one variable on the percentage of change in another variable. For example, an expenditure-to-growth elasticity would measure the effect of a 1 percent change in the stock of investments (or expenditures) on the change in growth outcomes at either the sector or the economywide level. This becomes useful for estimating the requirements for future growth in public investments to generate desired sector and economic growth targets. Using estimates of both expenditure-to-growth elasticity and growth-to-poverty elasticity, one can then estimate the level of resources required to achieve desired growth and poverty goals (Fan et al. 2008).

Are there any applications to draw from?

The early work undertaken by Fan, Zhang, and Rao (2004) for Uganda is a good example. Their study shows that the returns to past public investments were particularly high for agricultural research and development (R&D), rural feeder roads, and education (see Table 2.1). For every shilling invested in agricultural R&D, another 12.4 shillings were generated due to increased agricultural incomes. More recently, a study by Fan et al. (2008) used a simple elasticity method based on a number of econometric studies to estimate the level of resources required among 30 African countries for them to achieve MDG1. Other country case studies in Africa have used similar approaches in estimating resource requirements for meeting the CAADP growth target of 6 percent (Thurlow et al. 2008a, 2008b; Benin, Johnson, and Omilola 2010).

WHAT HAVE BEEN THE LESSONS AND EFFECTS OF ARD STRATEGIES ON OUTCOMES AND IMPACTS?

Why is this question important?

A critical part of any ARD strategy is to be able to determine at some point whether the strategy is on track to achieve its goals or whether at the end of its life people's lives will have been positively or negatively affected. Such a strategy helps to justify resource investments and ensure accountability. It also provides the option to adjust the strategy as new evidence becomes available on what has (or has not) been working. From the perspective of the goals of SAKSS, therefore, helping to provide credible and relevant analysis and information related to M&E is a fundamental prerequisite to promoting evidence-based policymaking. And although the importance of M&E systems is well documented (see, for example, Dalal-Clayton and Bass 2002 and Mackay 2007), they are often the most difficult part to set up and maintain. This is because ARD strategies operate within a broader complex environment that is inherently dynamic with respect to constantly changing domestic and global economic conditions, social and political trends, climatic shocks, and processes associated with implementing ARD strategies.

We now discuss the corresponding questions for strategy within the context of M&E, analytical methods used in monitoring and evaluating progress and impact, and examples of application. Guidance on the practical elements to consider in setting up an M&E system is discussed in the seventh section of Appendix A.

Table 2.1. Returns to government investment in rural Uganda

Investment	Central	East	North	West	All Uganda
Benefit–cost ratio					
Agricultural R&D	12.49	10.77	11.77	14.74	12.38
Education	2.05	3.51	2.10	3.80	2.72
Feeder roads	6.03	8.74	4.88	9.19	7.16
Murram roads	n.s.	n.s.	n.s.	n.s.	n.s.
Tarmac roads	n.s.	n.s.	n.s.	n.s.	n.s.
Health	1.37	0.92	0.37	0.96	0.90
Reduction in number of poor people per million shillings					
Agricultural R&D	30.23	92.15	243.92	67.97	81.14
Education	3.57	21.60	31.38	12.62	12.81
Feeder roads	10.51	53.85	72.82	30.49	33.77
Murram roads	4.08	11.88	14.80	9.77	9.70
Tarmac roads	2.59	13.12	62.92	9.39	9.73
Health	2.60	6.15	5.95	3.46	4.60

Source: Fan, Zhang, and Rao (2004).

Note: R&D means research and development; n.s. means that the respective coefficient was not statistically significant.

What are some corresponding questions for strategy?

An M&E system seeks to answer questions that determine whether development objectives are being fulfilled and if there is impact in order to help revise and improve future strategies, such as these:

- Are the allocation and level of inputs (for example, spending, investments, policy interventions) of the ARD strategy (or project) on target?
- How can the impact of these investments be traced to improvements in outputs (for example, productivity and the viability of production systems, food processors, agroindustries, markets, and trade)?
- Have these improvements affected outcomes (for example, incomes and the poverty status of target populations)?
- What factors have shaped (positively and negatively) the level of impacts achieved to date? What needs to be altered?
- What was the distribution of these intermediate impacts (for example, did they affect smallholders, equity, or gender or have other spillover impacts)?
- Can key ingredients of success or failure be identified based on past experience and lessons learned? What have been the roles of the public versus the private sector?

A desirable M&E system, therefore, is one that can not only monitor progress among key inputs, outputs, and outcome indicators but also evaluate the magnitude and distribution of impacts. In terms of progress, simple descriptive narratives of trends among the indicators help to answer the adequacy question: "Have expectations in terms of investment flows and achievement of the growth and poverty reduction targets been met so far?" However, it does not answer the key questions regarding hypotheses: "How effectively have different types of policies and investments addressed the goals so far?" and "What factors have shaped the level of impact that has been achieved?" These last two questions are more concerned with the evaluation part of M&E. Addressing all three provides key information that can help guide decisions on what to adjust as well as the scale and mix of investment priorities needed in order to keep a country's ARD strategy on track. In this way, it is a critical component of the strategic analysis agenda of a SAKSS, coming full circle to inform and strengthen future design and implementation of ARD strategies.

What is needed to answer these questions?

One of the first challenges in answering these questions is to accumulate sufficient reliable quality data from national statistical systems or project-level surveys from which to gather baseline information and set future targets. A second is to have an adequate M&E framework in place that defines a clear goal for the strategy or project; describes the interrelationships or causality among inputs, outputs, and outcomes; and defines a minimum set of indicators to monitor and evaluate. A third challenge, one integral to the development of the M&E framework, is to introduce sufficiently robust methods and tools for monitoring and evaluating impact. We now review these in more detail.

Availability of reliable data

The poor availability of reliable data in most African countries makes it particularly challenging to set up a national M&E system, just as it is to set up the other types of strategic analyses discussed in this chapter. Especially problematic are the frequency and quality of data collection on agricultural production and marketing. Most African countries have undertaken only a handful of these surveys since the 1970s. In Ghana, for example, the last comprehensive agricultural census was carried out in the early 1970s, followed by a smaller sampled survey in 1986.[7] Other socioeconomic surveys, such as the living standards measurement surveys, population and housing censuses, and demographic and health surveys, have been conducted more frequently, but these vary across countries. Even if data are available, the quality is often poor due to the overall weak technical and managerial capacities of the national statistical systems in Africa. Explanations for this have included a generally low regard for statistical information by policymakers, poor links among statistical systems and policy processes, inadequate government spending and technical assistance for statistics over long periods of time, and outdated statistical systems and legislation (Wingfield-Digby 2007; Kiregyera 2008).

Another weakness is the lack of timely and sufficient information on the flows of public-sector expenditures and investments in ARD, especially from public expenditure tracking surveys (Dorotinsky and Floyd 2004). Such information can serve as key input in setting baselines and targets for investment spending and in undertaking impact studies later on, in addition to helping improve accountability and public-sector management more generally. Typically much of this information is found within finance ministries, with details of agricultural expenditures available from sector ministries.

In many African countries there are increasing efforts to improve the data systems for agriculture in general (Kiregyera 2008).[8] One effort being led by the World Bank is the Living Standards Measurement Study and Integrated Surveys on Agriculture to improve household-level panel data on agriculture in Sub-Saharan Africa. According to the World Bank's website on this initiative, six African countries will initially benefit.[9] In time, and once the initiative has been scaled out to other countries, it should help improve and complement existing datasets for purposes of monitoring the performance and impact of ARD strategies in Africa. Another effort is the introduction of the African Union (AU) / NEPAD budgetary tracking system for agricultural expenditures to monitor national compliance with the Maputo Declaration.[10] This initiative has begun to make information on total agriculture spending available (AU/NEPAD 2005). However, because it does not cover disaggregate flows by subsector and type of investment, it is less useful for impact evaluation (Benin, Johnson, and Omilola 2010).

Developing an M&E framework for analysis

Simply having sufficient access to data does not guarantee a good M&E system. It also requires a credible M&E framework, as well as the tools to monitor progress and assess impact. As a start, either a logical or a theoretical framework can be useful in this regard. Both seek to lay out a simple structure that describes the causal relationships among inputs (for example, investment flows), outputs (for example, productivity), and outcomes (for example, growth and poverty), from which critical corresponding indicators can be identified for the purpose of monitoring and evaluating impact. Next we review a number of approaches under each of these broad frameworks, including their strengths and weaknesses. Readers are also encouraged to see the seventh section of Appendix A, which provides a practical overview of how to go about establishing an M&E system using either of these frameworks.

---

[7] Based on a preliminary appraisal of the state of agricultural data systems in Ghana conducted by IFPRI as part of the early phase of setting up a country SAKSS (see Chapter 5).

[8] Examples include the Marrakech Action Plan for Statistics (MAPS); the General Data Dissemination System of the International Monetary Fund; the global Partnerships in Statistics for Development in the 21st Century; the Reference Regional Strategic Framework for Statistical Capacity Building in Africa, a regional variant of the MAPS; the National Strategies for the Development of Statistics; and AFRISTAT, limited mostly to West Africa.

[9] A permanent link to this site is <http://go.worldbank.org/EG1SDRFER0> (accessed July 28, 2009).

[10] Signed in 2003, the Maputo Declaration directed all African Union (AU) member countries to increase investment in the agriculture sector to at least 10 percent of their national budgets by 2008.

Logical framework (logframe) approaches. The logical framework (or logframe) approach helps describe a simple flow chart of how inputs will achieve intermediate and final outcomes. Performance indicators are selected as part of a results-oriented logframe matrix measuring performance in terms of input delivery, implementing activities, producing outputs, and achieving targeted outcomes (Crawford and Bryce 2003; Kusek and Rist 2004; World Bank 2004a). The measurement of outputs can also include specifying the extent of coverage among target groups, including nontarget groups if spillovers are expected and measurable. Performance indicators are then selected at each stage along a simple causality chain, with assumptions about associated risks and other confounding factors that can influence performance and outcomes.[11] Adaptations to improve the basic logframe have occurred over time, leading to what Sartorius (1996) refers to in the title of an article as "third generation" logframes. Among these are team-based (or participatory) approaches, application of the logframe to research projects, development of software systems to support the basic logframe, better integration of the logframe with other project implementation tools, improved understanding and selection of practical indicators, and development of a better understanding of the logframe's use within institutions.[12]

Despite various adaptations of the logframe, its fundamental structure and utility as a management tool have not changed. The approach remains popular among institutions for tracking program performance and results over time. It also has the added advantage that it can be developed in a participatory fashion involving broader stakeholder groups. Its main drawback is that it is not reliable for undertaking credible impact evaluations because it is too static and rigid and relies on too many assumptions when leaping from inputs to outputs and on to outcomes based on simple linear causal relationships (Crawford and Bryce 2003; World Bank 2004a). In reality, the causality chain is far more complex. Many other factors influence outputs and outcomes along the chain and through time, including cross-sector linkages and multiplier effects. Alternatively, theoretical framework approaches try to more explicitly deal with these challenges.

Theoretical framework approaches and others. Theoretical frameworks seek to introduce more theory and analytical rigor along the entire length of the causal chain (Spinger-Heinze et al. 2003; World Bank 2004a; White 2006). One such approach is *impact chain analysis,* which essentially maps out the links in the causal chain based on underlying economic theory and evidence. While still based on the logframe, it allows for a more in-depth theoretical understanding of the cause-and-effect relationships or impact pathways among variables and the confounding effects of other potentially influencing factors. And as long as there are sufficient data, empirical evaluations of impact using econometric techniques can be modeled, as discussed further later. The main disadvantage of this technique is that it can easily become unnecessarily complex, data intensive, and costly.

An alternative to impact chain analysis is *outcome mapping,* which is more participatory and qualitative in nature, focusing on changes in development processes and outcomes (Earl, Carden, and Smutylo 2001; Smutylo 2005). Essentially it identifies expected impacts and outcome indicators based on a map of interrelated factors from intervention to outcomes within a group session. Given the complexity of considering many other noneconomic factors that can influence outcomes, this approach adopts visualization techniques to encourage participants to think through the different factors, their direct and indirect effects, and causal linkages. This is especially useful for tracing qualitative changes in development processes that are not easily measurable (for example, human behavior). A disadvantage of this approach is that it is limited by the complexity of causal relationships that arise as more factors and stakeholders are identified in the process.

More recently, the *impact pathway* approach was introduced in an attempt to deal with some of the shortcomings of outcome mapping and impact chain analysis (Douthwaite et al. 2003; Spinger-Heinze et al. 2003). This approach recognizes the presence of a number of impact chains and sequences in explaining the overall change process. The analysis of multiple impact pathways can be quite useful for impact evaluation because it avoids the pitfall of assuming a simple linear relationship between an investment and an outcome. For example, the analysis of development domains, discussed earlier in this chapter, recognizes the existence of multiple but distinct impact pathways. This is because how investments ultimately lead to outcomes depends on the types of combinations among multiple factors, such as agricultural potential, market access, and population pressure (Pender et al. 2001; Spinger-Heinze et al. 2003). The impact pathway approach, nonetheless, is not without the same drawbacks as outcome mapping and impact chain analysis. It can easily become unnecessarily complex, data intensive, and costly, especially when too many other factors and distinct pathways or outcome mappings are introduced.

No matter which approach is adopted, however, the real challenge from a practical perspective is to maintain sound theory and rigor while at the same time limiting the degree of complexity in drawing out the causality chain and in selecting

[11] A number of useful sources on how to go about selecting indicators and developing an M&E system, either at the macro or the project level, can be found in Dalal-Clayton and Bass (2002), IFAD (2002), Kusek and Rist (2004), World Bank (2004a), and McKay (2007).

[12] Maredia (2009) discusses how development paradigm shifts over time influenced the various adaptations to the logframe.

a minimum set of indicators for which reliable data exist. Depending on the strategy goals and underlying programs, a balance needs to be struck between the need to attribute impacts to program interventions and having in place a cost-efficient M&E system, one that has the least likelihood of measurement errors and is simple enough to allow interpretation of the information that comes out of it (see seventh section of Appendix A).

Evaluating impact

Developing an M&E framework for analysis is particularly useful in providing a simple structure for what is otherwise a complex system. However, effective evaluation of the impact of a strategy (or its underlying projects) after it has been in place for some time requires additional tools, both quantitative and qualitative. A number of quantitative tools exist for analyzing impact at the sector or the economywide level and at the project level.

Sector or economywide level. Evaluation of the sector or economywide impacts of public-sector interventions is particularly useful when the goal is to assess effects on aggregate welfare outcomes (for example, poverty) and their distribution. A number of approaches and tools can be used depending on the availability and quality of data. Where there is sufficient information on past investments and other key factors that influence outcomes, econometric and statistical methods can help test for the contribution of past changes in investments (for example, in agricultural research, infrastructure, health, and education) on changes in outcome variables. This is very much the same approach described in "How should resources be mobilized and allocated across the different economic sectors and geographic regions?" which emphasized that ex post impact evaluation is particularly relevant for informing the design of future strategies. Some examples include the work by Fan, Zhang, and Rao (2004) for a broad category of investments. For agricultural R&D, see the reviews of economic rates of return by Echeverria (1990); Masters, Bedingar, and Oehmke (1998); and Alston et al. (2000).

Given the poor availability and quality of data in many African countries, econometric techniques may not be feasible. Under these conditions, simulation and programming models are useful alternatives. For agricultural R&D, for example, economic surplus models (such as the DREAM model discussed in the sixth section of Appendix A) can be used to estimate economic rates of return to investment as long as there is sufficient information on key technology and behavioral parameters (Alston, Norton, and Pardey 1995). Additionally, programming techniques can be used to estimate changes in agricultural performance (as measured by total factor productivity) that is due to technical change and thus past investments in R&D (Farrington, Thirtle, and Henderson 1997). For broader categories of investment, it is also possible to apply economywide simulation models as in "What are the economywide options for reaching high-end ARD goals?" if such models already exist. This can be particularly useful for capturing the impact of broader sectorwide investments on overall economic growth. If economic models are not feasible, simpler elasticity models can also be considered as suggested in Box 2.4.

The application of economic simulation models to the evaluation of impact after the fact (or ex post) essentially involves simulating how much an actual change in investments or policy may have affected outcomes. A number of indicators that serve as input into the models can be monitored periodically to assess their potential impact on outcomes. As data on outcome variables become available, model results can be compared with actual values. If the model predicts an outcome that is below or above the actual observed outcome, it is possible that a number of confounding factors could explain the difference if data and model specification errors are assumed to be minimal. Results can also be compared against a baseline scenario in which the intervention is removed to describe a situation "with and without" as a type of counterfactual analysis (see the example of Bell, Hazell, and Slade 1982).

An emerging technique is the use of spatial analysis tools for impact assessment. This has become possible as GIS tools, satellite imagery, and computer hardware and software have advanced over the years. As a result, for example, an increasing number of countries are able to produce high-resolution poverty maps. Maps of this kind, which show disparities in poverty across geographic space, including associated livelihoods and assets, are not only relevant to targeting future investments (as discussed in "How can ARD strategies be targeted to address the diversity of opportunities and challenges within a country?') but can also serve as a monitoring and evaluation tool. Statistical techniques can be applied where there are sufficient data to be able to associate a number of key geographic and socioeconomic factors with changes in welfare over time (see, for example, Pender et al. 2001 and Minot and Baulch 2005).

Impact evaluation at the sector or the economywide level has its limitations. The links between investments and outcomes are typically too aggregate to be linked back to any particular intervention (Maredia 2009). The common problem of insufficient data makes econometric and spatial analysis difficult and often impractical. Although economic simulation models are more feasible under these conditions, their accuracy depends on how well the underlying data, model specification, and behavioral assumptions represent the real world. In many instances, behavioral parameter estimates in a model are drawn from different periods in time and/or different locations.

Project level. A more detailed and selective impact evaluation at the individual project level is often desirable because

Box 2.4: A simple elasticity model for assessing impact and attribution

Impact can also be assessed using available econometric estimates on the behavioral relationships among key variables. For illustrative purposes only, suppose there are known estimates of the marginal impact of certain investments on poverty (in terms of elasticities) and we also know the following causal relationships among them:

$$\dot{POV} = 0.8\dot{ROADS} - 1.1\dot{RDE} - 1.5\dot{MKT},$$

where

- RDE = investments in R&D and extension (science and technology),
- ROADS = investments in rural roads (infrastructure),
- MKT = investments in market institutions/policy reforms (markets), and
- POV = number of people earning less than US$1/day or number of people falling below the national poverty line.

and the dot above each variable refers to a percentage of change (for example, $POV = dPOV/POV$).

Further, suppose we observe the following changes in the key variables between two periods:

$$\dot{POV} = 5.2\%, \dot{ROADS} = 1\%, \dot{RDE} - 1\%, \dot{MKT} - 2\%.$$

Then the predicted change in poverty is –4.9 percent. Because the actual change in poverty is –5.2 percent, the difference of 0.3 percent is attributable to a residual variable. The reduction in poverty can now be attributed to its various sources. For example, the share due to change in ROADS is calculated as (–0.8)(1.0)/(5.2) × 100 percent = 15.4 percent. That is, of the observed 5.2 percent reduction in poverty, 15.4 percent can be attributed to increased roads. If the share of the increase in roads due to investments is known (say 0.3), then multiplying 15.4 percent by this share will give the percentage decline in poverty due to investments in roads (in this case 15.4 × 0.3 = 4.6 percent).

Similar calculations show that changes in RDE account for 21.2 percent of the reduction in poverty, changes in MKT account for 57.7 percent, and the residual accounts for 5.7 percent. Again, if the shares of the changes in RDE and MKT due to investments are known, the predicted change in poverty is –4.9 percent. Since the actual change in poverty is –5.2 percent, the difference of 0.3 percent is attributable to a residual variable.

Source: Johnson and Resnick (2004, Box 12).

it can directly test for attribution. At this level of analysis, experimental or randomized approaches are potentially more feasible to employ. A key advantage of doing so is that randomization allows measurement of impact against some counterfactual while guarding against problems of selection bias in the evaluation. In other words, it can answer questions about how individuals who participated in a program would have fared in the absence of the program or how those who were not exposed to the program would have fared in the presence of the program (White 2006; Duflo, Glennerster, and Kremer 2008; Maredia 2009). In this way it can explicitly address the question of causality and reveal the conditions under which changes in the welfare of the population have been brought about by, or can be attributed to, specific interventions. To ensure that sufficient information is collected for randomization, however, early preparations and resources for monitoring and evaluation must be in place when the project is in its design phase to provide "before-and-after" and/or "with-and-without" intervention comparisons.

The experimental or randomization approach is not without its critics, however, who question its usefulness for informing policy and strategy design altogether (Deaton 2009; Ravallion 2009). An obvious limitation is that typically the desire to undertake an evaluation arises when a project has already ended, which makes the condition for randomization impossible. Another is the ethical question of excluding a control group from treatment when it involves welfare-improving interventions. Deaton (2009) cites a few other problems, including the limitations of scaling up to general equilibrium effects that occur at higher levels of impact but are unaccounted for at the project level, the existence of uniquely defined local conditions that affect outcomes but are not always present in other locations, and the observation that simply identifying what worked in a project is not as useful for policy if it does not also show why it worked. Finally, the high costs of this approach are not always justifiable if it does not contribute much to policy considerations. According to the Independent Evaluation Group, an independent unit within the World Bank that is tasked with evaluating what works and what does not among Bank projects, costs for experimental impact evaluations can range between US$300,000 and US$500,000 per study (White 2006). Therefore, the cost for experimental impact evaluations relative to the information expected from them should always be accounted for before undertaking such case studies.

Other more commonly accepted nonexperimental and practical alternatives are available, however. Among them are statistical and econometric techniques that compare outcomes between project participants and nonparticipants within a target population, such as controlling for observables, performing regression discontinuity design estimates, or using difference-in-differences or fixed-effects approaches, as well as using instrumental variables (Duflo, Glennerster, and Kremer 2008; Maredia 2009). These do not face the inherent problems and high costs associated with the experimental methods. Another valuable alternative is the traditional social cost–benefit analysis (or rate-of-return studies) using the economic surplus method, but this works only if there is sufficient underlying information on costs and behavioral assumptions associated with the project (Gittinger 1984; Alston, Norton and Pardey 1995; Masters, Bedingar, and Oehmke 1998; and Alston et al. 2000).

The choice of economic tools ultimately depends on many factors: whether impact is being evaluated at the project, sector, or economywide level; the type of questions being asked to ensure policy relevance; data availability and type; pre-existing models; the resources and time available for analysis; and the capacity or skills of the evaluator. Ravallion (2008) offers some useful steps for evaluators to consider in selecting the most appropriate methods and approaches.

Incorporating qualitative assessments

Sometimes impact evaluations are not amenable to a quantitative set of economic analyses, such as when a project mostly provides a service or has a strong social dimension. Nor are they complete in explaining impact without considering other critical social and political explanations. Qualitative analyses can help provide vital information about the social and political context, including processes, under which a project's impact is being evaluated. Many of these factors can influence outcomes during the design and implementation of an ARD strategy by affecting the efficiency and effectiveness of program interventions and projects. In this regard, the *impact pathway* approach discussed earlier is a good example of such a qualitative assessment because it pays attention to understanding processes and behavior in explaining impact. Increasingly, participatory approaches to impact assessment such as these have become attractive in the development community because they are more flexible, can use rapid appraisal techniques (for example, such as civil-society report cards), and provide a better understanding of the perceptions and needs of targeted beneficiaries than can other methods (White 2006; Maredia 2009). Other methods include the sustainable livelihoods framework, which considers a broader set of social and political explanations in addition to the economic ones in assessing a project's impact on welfare outcomes (Adato and Meinzen-Dick 2007). A primary focus of the framework is to draw causal links among a household's access to broadly defined assets, its chosen livelihood strategy, and welfare outcomes. The range of assets goes beyond the economic ones (natural resources, human capital, financial capital, and physical capital) to include social (for example, networks, safety nets, membership in organizations) and political (for example, citizenship, membership in political parties) assets.

Qualitative approaches to impact evaluation add important dimensions and insights that are quite complementary to traditional quantitative ones. They have the advantage of being open ended and able to start answering questions about why and how a strategy or project may have failed or succeeded rather than what caused the failure or success. Disadvantages include the subjective nature of gathering information, the lack of good statistical properties to validate results, and the fact that the information derived is not always transferrable (Maredia 2009).

Having reviewed the various approaches to developing an M&E framework, data needs, and a range of impact evaluation tools, the choice of which to use will always depend on the type of project or ARD strategy; the questions being asked;

the level of complexity of the known causal relationships; data availability, frequency, and type; existing tools; the skills of the analysts involved; the budget and costs for M&E; the time horizon for analysis; and country circumstances and processes. With this knowledge, decisionmakers, technical analysts, and the key stakeholders involved must together decide which approach and what tools provide the most robust and cost-effective M&E for the particular project or ARD strategy in mind (see seventh section of Appendix A for some useful guidelines).

Are there applications to draw from?

Given the infrequent data collection and availability for M&E systems in Africa, various methods and initiatives have been introduced to improve M&E data systems in general at country, regional, and global levels. Because we discussed the various initiatives for improving national statistics for monitoring and evaluation earlier (see "Availability of reliable data"), we focus on a few other examples here.

At the country level, finding alternatives for dealing with the problem of infrequent data collection on outcome indicators is not uncommon. In Kenya, for example, an annual income proxy measurement was developed to monitor the impact of projects funded by the United States Agency for International Development. The proxy was generated using econometric techniques to associate past actual income changes with a number of variables for which data are more frequently available (Tschirley and Mathenge 2003). This type of work makes a strong link between the types of economic activities that target populations are engaged in and the incomes they derive from these activities. The use of such income proxies can allow for more frequent and lower-cost monitoring and impact evaluation. At the regional level, efforts have been under way to harmonize and standardize data systems on agricultural expenditures and other key indicators. The AU/NEPAD expenditure tracking system and CAADP M&E by ReSAKSS (AU/NEPAD 2005; Benin, Johnson, and Omilola 2010) are useful examples.

At the global level, efforts to improve the harmonization of M&E approaches and standards are being led by the United Nations in the Partnership in Statistics for Development in the 21st Century (PARIS21, www.paris21.org). Founded in November 1999 by the UN, EC, Organization for Economic Co-operation and Development (OECD), International Monetary Fund, and World Bank, the initiative is working to encourage and assist low-income countries to improve their national statistics. Among donors, a notable commitment to improved harmonization of M&E systems is underscored in the Paris Declaration of 2005 on the effectiveness of aid. The declaration commits OECD member countries and their institutions to measurable actions in ways that ensure that their actions are more harmonized, transparent, and collectively effective. One important effort that has appeared as a result is the establishment of the Global Donor Platform for Rural Development, which promotes greater coordination, communication, and shared learning among donors in the effort to support ARD in general and CAADP in particular (GDPRD 2006). It does this by formulating common donor policies, reducing the duplication of effort, and improving mutual understanding among all partners and advising partners as to how harmonization and alignment can make ARD initiatives more effective (see www.donorplatform.org).

Among the M&E frameworks, the logframe is one of the most widely used approaches to the results management of projects and strategies among institutions, governments, and development agencies. The strategic objective framework of a SAKSS shown in Figure 4.2 and the corresponding logframe in Table 4.1 illustrate its usefulness in managing a SAKSS network strategy. Its application among African governments is reviewed by Mackay (2007). An example for Uganda is also summarized in Box 2.5. On the theoretical framework, the proposal of a CAADP M&E system by Benin, Johnson, and Omilola (2010) adopts an impact chain analysis approach to describe the relationships among CAADP's investment pillars and poverty outcomes (see Figure 2.2). Associated indicators, including potential confounding factors, are then selected based on this framework.[13] Simple trends analyses can also be particularly useful during implementation of the CAADP to track the progress of some of the key input, output, and outcome indicators (see the example in Box 1.3 and Figure 3.2).

For impact evaluation at the sector level, aside from the case studies noted in the corresponding previous section, a number of studies assessing the impact of international agricultural research can be found in Evenson and Gollin (2003). At the project level, examples of the application of the experimental approach to impact evaluation are reviewed extensively by White (2006). A well-known example, nevertheless, is the randomized evaluation of the PROGRESA project in Mexico (Hoddinott and Skoufias 2003). For nonexperimental methods, see the work by Benin et al. (2007) in assessing the impact of a new public–private extension program in Uganda.

Case study applications of impact evaluation using both qualitative and quantitative methods have been particularly useful in drawing lessons and understanding the context and conditions for success. For example, the recent work by Resnick and Birner (2008) surveys four case studies in West Africa on the formulation of ARD strategies and the role of participation. Although the authors found no lack of participation, they highlight problems associated with false expectations about the effect of participation on policy outcomes and the failure to link strategies resulting from participatory processes with actual

[13] See Benin, Johnson, and Omilola (2010) for details on a list of proposed indicators and associated impact evaluation tools linked to the overall M&E framework.

Box 2.5: The experience with monitoring and evaluation systems in Africa for PRSPs

Drawing on the experience of poverty reduction strategy papers (PRSPs), a recent study by the Overseas Development Institute (ODI) reviewed the experience with establishing monitoring and evaluation (M&E) systems for the PRS (ODI 2003). The PRSPs put a focus on achieving results, such as the Millennium Development Goals (MDGs). As a result, there has been an increasing focus on improving national statistics in order to have a well-functioning M&E system, although some have argued that the emphasis on improving national monitoring systems has been driven principally by donor requirements for data and reporting (World Bank 2004b). According to the ODI study, although African governments have paid more attention to monitoring higher-end goals, such as tracking budget allocations and assessing progress against growth targets and the MDGs, little has been done in the middle, which refers to the intermediate steps in the causality chain. The study cites the work of Booth and Lucas for Uganda (2002). Additionally, there is a dearth of any serious efforts to evaluate the link between investments and actual results on the ground.

A principal challenge among even the African countries better equipped for M&E is rationalizing the system so it can be more effective. Problems of too much data and little relevant information are especially typical. In response, the Government of Uganda, for example, has created a National Integrated M&E System (NIMES) to serve as an umbrella M&E system within which existing systems will be coordinated and harmonized. According to Mackay (2007), NIMES has helped reduce what was originally a very large number of performance indicators, focusing on outputs, outcomes, and impacts at the sector level.

For most other African countries, the generally weak demand by governments for M&E information, poor national data systems, and poor economic analysis and statistical capacities all contribute to the absence of reliable M&E systems. Mackay (2007, 48), drawing on World Bank (2004a), suggests a number of important actions that could be taken at a minimum to improve M&E:

- Improving the public financial management systems for tracking spending
- Conducting public expenditure tracking surveys
- Conducting surveys of satisfaction and perceptions of the quality of government services
- Conducting rapid appraisals of "problem" projects or programs
- Establishing national and sector statistical collections, especially around the MDGs
- Compiling sector ministries' administrative data

Sources: Booth and Lucas (2002), ODI (2003), World Bank (2004a, 2004b), Mackay (2007).

decisionmaking. A case study review of successes in African agriculture by Gabre-Madhin and Haggblade (2004) and also summarized in Haggblade (2004) is especially noteworthy. The case studies provide a rich source of experiences of public-sector interventions across a number of agricultural subsectors and development themes. Based on their review, the most successful interventions were found to be those related to soil and water conservation, replication of proven commodity-specific breeding and processing successes (for example, for cassava), marketing and information systems, vertical supply chains to improve efficiency, and improving regional cooperation in trade and agricultural technology. The more recent global review of evidence of successes from past ARD interventions by Spielman and Pandya-Lorch (2009) is also an excellent source.

Conclusion

SAKSS has been conceived as a network in which data, tools, and knowledge at the micro, meso, and macro levels are compiled, analyzed, and disseminated to inform the design, implementation,

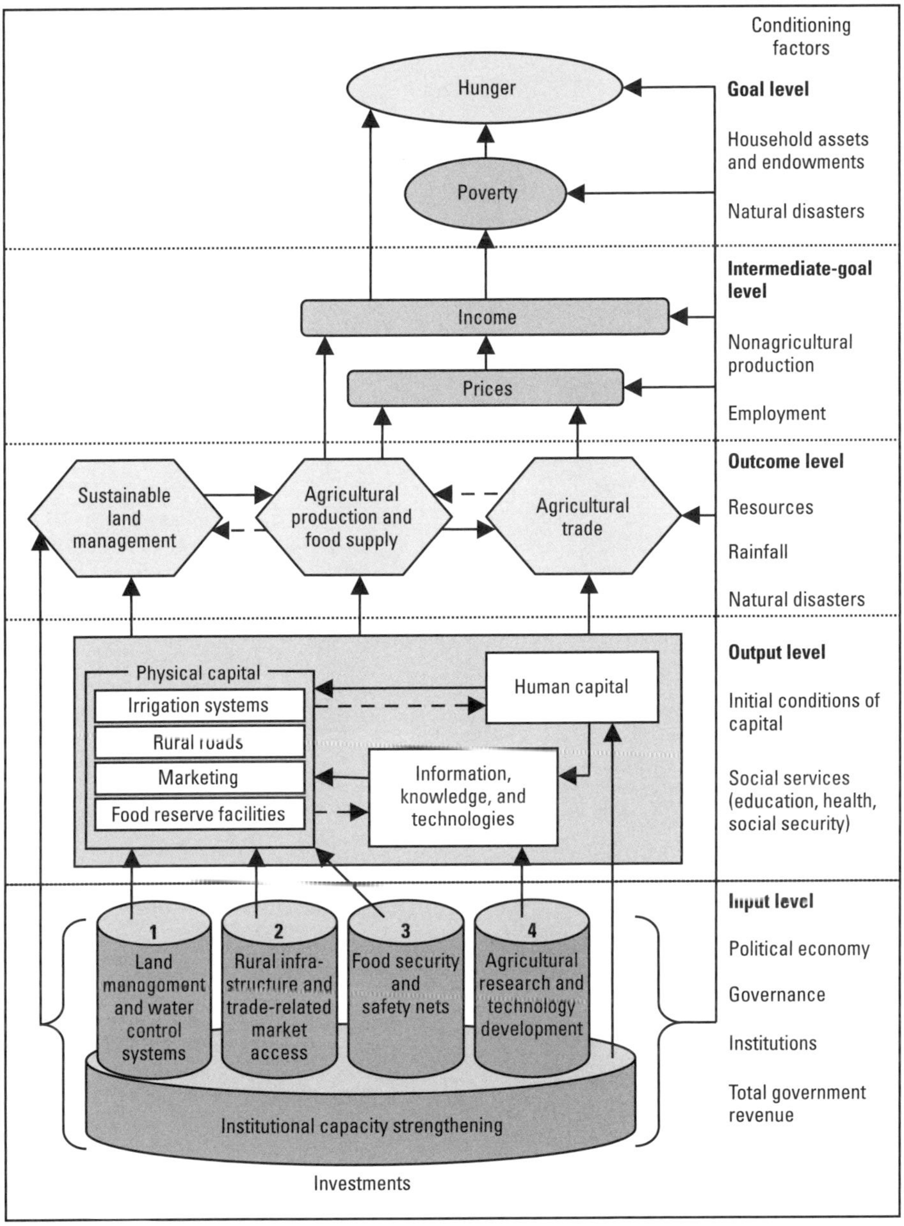

Figure 2.2. A theoretical framework of the Comprehensive Africa Agriculture Development Programme strategy for monitoring and evaluation

Source: Benin, Johnson, and Omilola (2010, Figure 1).

and M&E of ARD strategies. Economywide models, impact assessment tools, and M&E approaches are therefore critical features of the strategic analysis undertaken by a SAKSS. Because of the inherently spatial characteristics of agriculture and rural economies, another distinguishing feature is the emphasis on spatially relevant information. This not only facilitates location-specific strategic analysis but also provides a basis for linking ARD with other aspects of rural development, including health, education, the environment, and safety net programs.

Significantly, SAKSS is not an analytical effort to revive the top-down development planning approaches that have prevailed in the past. Rather, as a network, SAKSS is intended to help foster and contribute to an appropriate balance between credible research and relevant stakeholder input engaged throughout the process of undertaking the strategic analysis described in this chapter. This is the purpose of the second component of a SAKSS, the "knowledge support system," which is discussed in Chapter 3.

Chapter 3: The "Knowledge Support System" Concept and Approach

The *knowledge support system* component of SAKSS is defined as a network that serves the evidence needs of ARD strategy formulation and implementation. This network includes individuals such as researchers, policymakers, and development practitioners and institutions such as government agencies, universities, research institutes, development organizations, and private-sector and civil-society groups. These individuals and institutions are linked through collaborative research and strategic analysis, capacity strengthening, and dialogue. Through these activities evidence is compiled, synthesized, and packaged on a timely and reliable basis so it can have direct relevance to the policy deliberations or processes that shape and guide ARD strategies. Within each country context it is unknown whether there is sufficient room to maneuver the local political process in order to bring evidence to bear during the formulation and implementation of ARD strategies.

THE RESEARCH-POLICY DIVIDE

How research or evidence feeds into the policy process in developing countries is not very well understood.[14] Yet its importance in helping to inform and formulate future policies is often advocated as critical given the observation that many developing countries rely very little on science-based evidence when making policy decisions (Juma and Clark 1995). There is a growing body of literature that theoretically and empirically seeks to explain this challenge of closing the research–policy divide or gap in Africa. The ongoing work by the International Development Research Centre (IDRC) in Canada and the Overseas Development Institute (ODI) in the United Kingdom is particularly noteworthy.[15] Another team of researchers from Harvard University takes a broader perspective on the effectiveness of knowledge systems for sustainable development, offering a few African examples (Cash et al. 2003).

A commonly accepted viewpoint in the literature is that policymaking in general is a dynamic and complex process. In Africa, for example, it has even been referred to as a chaos of purposes and accidents (Clay and Schaffer 1984). The process can involve multiple actors (individuals and organizations) and be defined by local political, social (cultural and belief systems), and institutional (bureaucratic structures and capacities) realities. And because it involves people, vested interests among a few powerful elites, corruption, and external influences can also play a distinctive role, as they still do in many African countries (Juma and Clark 1995). Power relations (people) and ideas (based on both tacit and explicit knowledge) are therefore particularly important. Scientific knowledge often plays only a marginal role in the decision-making process (Sabatier 2007).

Achieving a good understanding of a country's own policy process, no matter its shortcomings, is therefore an important first step in understanding how evidence-based information can play any particular role in it. The concern is not simply about how to improve the transfer of research into policy and vice versa but even more about understanding the peculiar conditions under which links between the two can be made more effective.[16] Theoretical explanations of the policymaking process can offer useful perspectives on and frameworks for embedding research in the policy process.

One useful comprehensive review by Keeley and Scoones (1999) describes at least three dominant views: (1) a linear, logical approach; (2) an iterative, muddling-through, incremental approach; and (3) an approach centered more on discourse. Underlying these perspectives are assumptions about the types of relationships among the actors involved—from government to the research community, the private sector, and civil society. The linear, logical approach is no longer considered valid because it assumes that policy decisions follow a simple linear process of planning, implementation, and evaluation. In reality, according to the iterative viewpoint, they typically undergo an incremental and complex process. The

[14] Throughout this technical guide we define *research* loosely as including data, knowledge, and ideas. At the same time, we also refer to *evidence and information,* which we use interchangeably with *research,* still referring to the same things: data, knowledge, and ideas. Nevertheless, some of these have their own distinctive definitions, which we summarize in Appendix E.

[15] For IDRC, see the recent work by Ayuk and Marouani (2007) and Carden (2009). For ODI, see Crewe and Young (2002); Young (2005); and Jones et al. (2009).

[16] A number of political and social analysis methods offer some possibilities for doing this (World Bank 2007a). Cash et al. (2003) use a case study approach to analyze these links.

discourse argument goes further by introducing the presence of dynamic processes involving policy networks, advocacy coalitions, and policy narratives (Neilson 2001; Sabatier 2007). Both approaches assume a more bottom–up, participatory policy process compared with the linear model, one more reminiscent of democratic or pluralistic societies. Other perspectives exist but more or less overlap one of these dominant views.[17]

From a practical perspective, there are many reasons that there are gaps between research and policy. Drawing from the work of Stone, Maxwell, and Keating (2001), for example, Livny, Mehendale, and Vanags (2006) summarize these in five main areas. The first occurs when there is a lag between the time that evidence appears and the time that it is seriously considered during policymaking deliberations. This happens, for example, when the validity of the research or information is still contested or when it generally performs an enlightenment function. This can occur due to the inherently divergent cultures of both groups, with policymakers demanding more certainty about research results than researchers can give. Another reason for gaps is attributed to a supply failure or a public goods problem, such as poor-quality data and research results. This is quite common in many African countries. A third is a simple communication failure. For example, policymakers and/or researchers may not be keenly aware of the policy relevance of the research, in part because of the inherently different cultures and professional incentives surrounding the two. Policymakers, for example, typically want very practical solutions to a current policy problem. In contrast, researchers are more interested in theoretical abstracts and take a long-term perspective. A fourth reason for gaps is associated with a demand failure. This is the case when policymakers are simply not interested in considering the evidence because they do not think it matters, they are incapable of understanding or using the research, or they simply distrust it. Distrust for policy research that has been conducted and facilitated by donors occurs often in developing countries. Omamo (2004) offers a useful narrative of this problem in Africa. A fifth reason is the control of research by a few powerful elites who do not pay attention to the real needs of the intended beneficiaries of the research or, worse, wish to censor the results.

Given these observations, a growing body of literature pays a lot of attention to understanding the types of linkages under which the gaps between research and policy can be more effectively narrowed. Most emphasize a participatory approach, finding that the closer the engagement among all actors involved—researchers, policymakers, and practitioners—throughout the process of generating and disseminating research results, the more likely it is that policy outcomes will be influenced by the research (see, for example, Cash et al. 2003). Recognizing the inherent weak link between researchers and policymakers, one set of literature examines institutional and network arrangements that serve important intermediary, boundary, or brokerage functions to bridge the research and policy divide (Cash 2000; Guston 2001; Jones et al. 2009).[18] In essence, the boundary function is critical in improving communications and links between the two. Other sets of literature adopt a network theory approach. The network here is defined as involving a range of policymakers, researchers, practitioners, and other actors who are all linked by shared interests and resource relationships. Such networks usually have strong incentives to influence policy outcomes and maintain the status quo (Howlett 2002). The goal of the network approach is to decipher the complex pattern of interactions in order to determine the existing structure of network linkages among individual actors and organizations that ultimately influences policy change (Howlett 2002; Mendizabal 2006). Another perspective derived from the agriculture research and extension literature pays more attention to the organizational structures of innovation, or "innovation diffusion," and how it affects ideas for research and knowledge and ultimately how it can be acted on by both researchers and policymakers (Hall et al. 2000; Hall, Mytelka, and Oyeyinka 2005). All are relevant in exploring the formal and informal institutional arrangements, linkages, and relationships among the actors involved in generating and bringing evidence to bear in the policy process.

Understanding the types of links and relationships associated with bringing evidence to bear in the policy process is very important but not sufficient on its own. It is just as important to ensure that the evidence generated is credible on the supply side. As pointed out in Chapter 2, a crucial problem in many African countries is the poor supply of and demand for quality data, statistics, and policy analysis. Simply having access to good national statistics can strengthen and promote evidence-based decisionmaking. It can help highlight emerging issues, forecast future trends, improve the targeting of interventions using spatial poverty maps based on household surveys and census data, monitor progress toward development outcomes, and ultimately promote more transparency and accountability in government (Scott 2005). On the demand side, the capacity of policymakers to comprehend and value research results and statistics matters just as much. It is no use spending significant resources on state-of-the-art national statistics if they are not going to be used by anyone. This quickly becomes a chicken-and-egg dilemma. Should governments invest more in gener-

[17] For a more comprehensive review of the literature, see Neilson (2001), Stone (2002), Omamo (2004), and Sabatier (2007).

[18] A special issue of *Science and Public Policy* (vol. 30, no. 4, August 2003) takes up this issue of boundary spanning for bridging research and policy.

ating quality data and statistics in the hope that policymakers will eventually use them? Or should they work on improving the capacities of policymakers to learn about the value of statistics? This dilemma creates a serious challenge for evidence-base policymaking in Africa.

Establishing stronger links among the different communities—researchers, analysts, and policymakers—in the policy process is one practical way to start addressing these capacity gaps. The goal of the knowledge support system component of SAKSS is to do just that, to strengthen the links and capacities for greater evidence-based dialogue and policymaking while also ensuring the quality of the evidence generated. Therefore, capacity strengthening and strategic analysis are important activities of SAKSS, in addition to promoting stronger links for policy dialogue.

The Knowledge Support System Framework of SAKSS

The term *knowledge support system* is derived from the more general concept of a knowledge system. Although the term knowledge system is often used ambiguously, encompassing a wide range of definitions, applications, and structural arrangements, it usually refers to the process of "effectively connecting those who know with those who need to know, and converting personal knowledge into organizational knowledge" (Creech and Willard 2001, 8). Therefore, how effectively these systems bring knowledge to bear in the policy process depends on how well individuals (both the actors who know and those who need to know) and the organizations they represent (for example, research institutions, nongovernmental organizations, private organizations, and government agencies) are linked to promote dialogue around the knowledge products (that is, information and results of research) and policy issues.[19]

Knowledge systems come in various formats and are designed for different purposes. For example, large institutions engage in knowledge management to improve the efficiency and effectiveness of their operations. Extension and market information systems for farmers and entrepreneurs represent another type of knowledge system. The Agricultural Knowledge and Information System of the World Bank and the Food and Agriculture Organization of the United Nations (FAO) is a good example (Berdegué and Escobar 2001). Other knowledge systems include Internet-based systems for improving the access of researchers and policymakers to high-quality, relevant, and timely agricultural information. There are quite a few examples out there. Access to Global Online Research in Agriculture provides free or low-cost access to major scientific journals in agriculture and related sciences to public institutions in developing countries.[20] The Global Information Early Warning System of FAO and the Famine Early Warning System Network funded by the United States Agency for International Development (USAID) are Internet-based networks that provide early warning systems against possible shortfalls in food supplies to prevent famines and allow policymakers to plan for both short- and long-term responses.[21] The Southern Africa Regional Poverty Network promotes debate and knowledge sharing on poverty reduction processes and experiences in Southern Africa.[22] Comparably, an online SAKSS is intended to provide access to key information and trends relevant to ARD in Africa. An excellent example is the ReSAKSS website, designed to provide key information on agriculture in support of the Africawide CAADP agenda for peer review. This is illustrated in Figure 3.1.[23]

Regardless of the type of knowledge system, most depend on formal or informal networks with other individuals and institutions. Networks can be broadly understood as a combination of people and/or institutions that are usually dispersed over wide geographic areas but are connected through appropriate communication technologies. Indeed, formal networks have evolved rapidly with the emergence of information and communications technologies (ICTs), which have enabled more and more people to become engaged around shared interests or communities of practice regardless of geographic location. In the arena of international development, the rise of networks has also been driven by the awareness that development issues are highly complex and cannot be solved by narrow approaches as well as by a belief that networks offer a better means for researchers to exchange information and ideas and interact with practitioners in order to affect public policy (Creech and Willard 2001). However, networks do not necessarily operate effectively to influence policy. Certain characteristics can improve or impede their ability to affect policy, including size, representation, quality, leadership, communication and ICT capabilities, organizational and management structure, and sustainability (Perkin and Court 2005).[24]

---

19 Cash et al. (2003) provide some useful real world examples.

20 See <http://www.aginternetwork.org/en/> (accessed July 28, 2009).

21 See <http://www.fao.org/GIEWS/english/index.htm and http://www.fews.net> (accessed July 28, 2009).

22 See <http://www.sarpn.org.za> (accessed July 28, 2009).

23 See <http://www.resakss.org> (accessed July 28, 2009).

24 Again the ReSAKSS application is a good example. It includes formal networks within each of the major subregions—Southern Africa, Eastern and Central Africa, and West Africa. The networks derive their membership from key individuals and groups associated with regional economic organizations and member countries, including research organizations, agriculture ministries, statistical bureaus, and farmer trade associations (this is covered in more detail in Chapter 5). In the absence of any formal external evaluation, it is still too early to tell how successful these networks have become.

Figure 3.1. ReSAKSS website, homepage

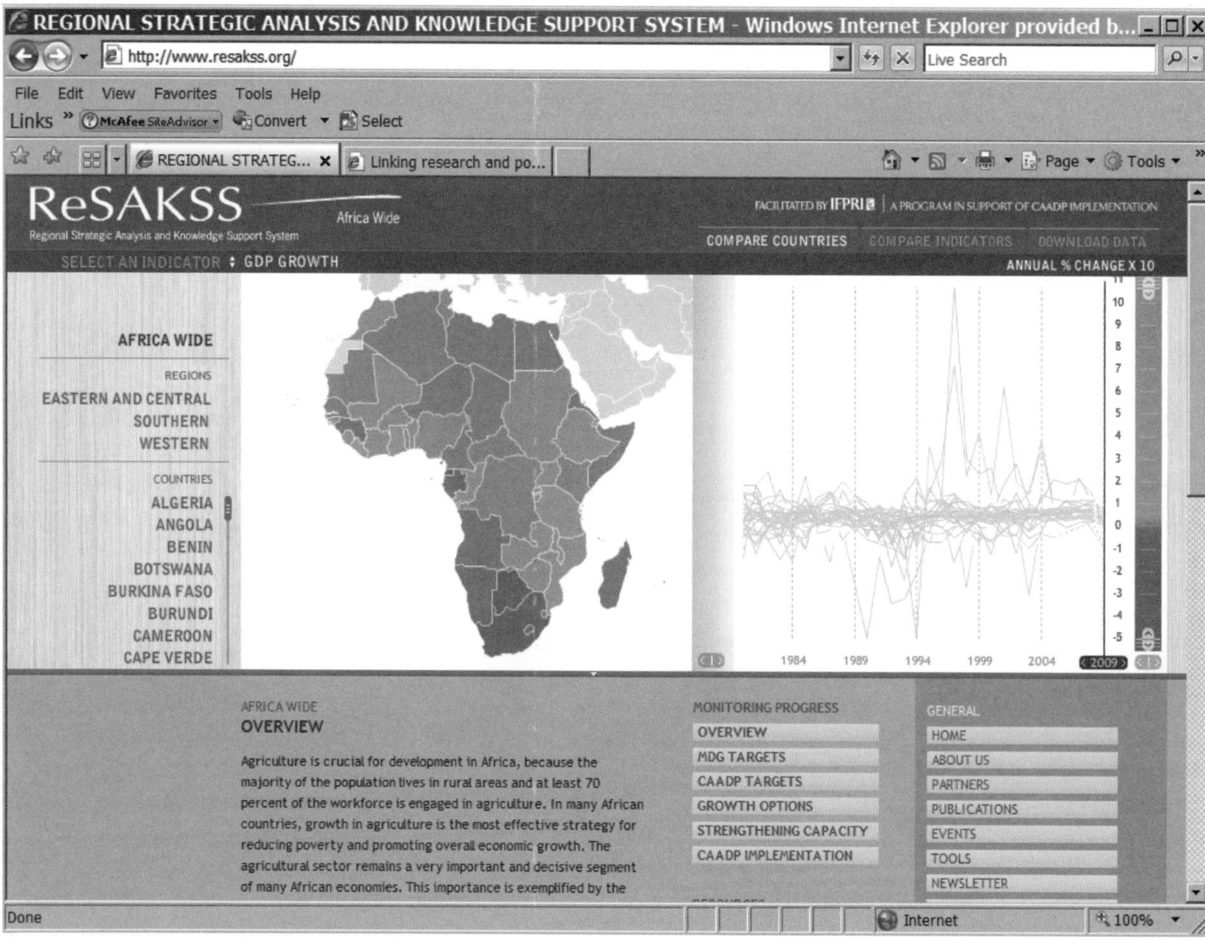

Source: Print image of ReSAKSS website homepage, <www.resakss.org> (accessed July 8, 2009).

How the SAKSS network, in particular, is intended to serve as a knowledge support system in the context of an ARD strategy is best illustrated using the Context Evidence and Links Framework developed by ODI and described by Crewe and Young (2002), Court and Young (2003), and Young (2005). The framework involves four key elements: external influence, political context, evidence, and links. It is appropriate for our purpose because it describes the complex interrelationships among a diverse group of actors given a local political context and external environment, and thus the manner in which evidence can play a role in contributing to policymaking. Nevertheless, it should be underscored that the complexities of the research–policy interface cannot be adequately represented in a single framework because it involves many other dimensions, such as the inherent dynamics of a particular sector regarding the supply of and demands for information (Jones et al. 2009). Therefore, the framework used here is intended for illustrative purposes only.

Figure 3.2 describes how a SAKSS network would fit within the ODI framework and how it could ultimately provide input into the ARD strategy development process of planning, implementation, and M&E. From the figure we see that the processes and outcomes of the planning, implementation, and M&E activities are themselves greatly influenced by the interrelationships among the three spheres in the center, including the external environment surrounding them. Leadership and governance at the top emphasize the principal role this plays in managing the ARD strategy process itself, which is inherently directly influenced by the political context and the external environment. Usually the leader and manager of the ARD strategy sits in the Ministry of Agriculture or another government agency or department charged with this responsibility.

Figure 3.2. The role of SAKSS as a "knowledge support system" for informing agricultural and rural development strategy

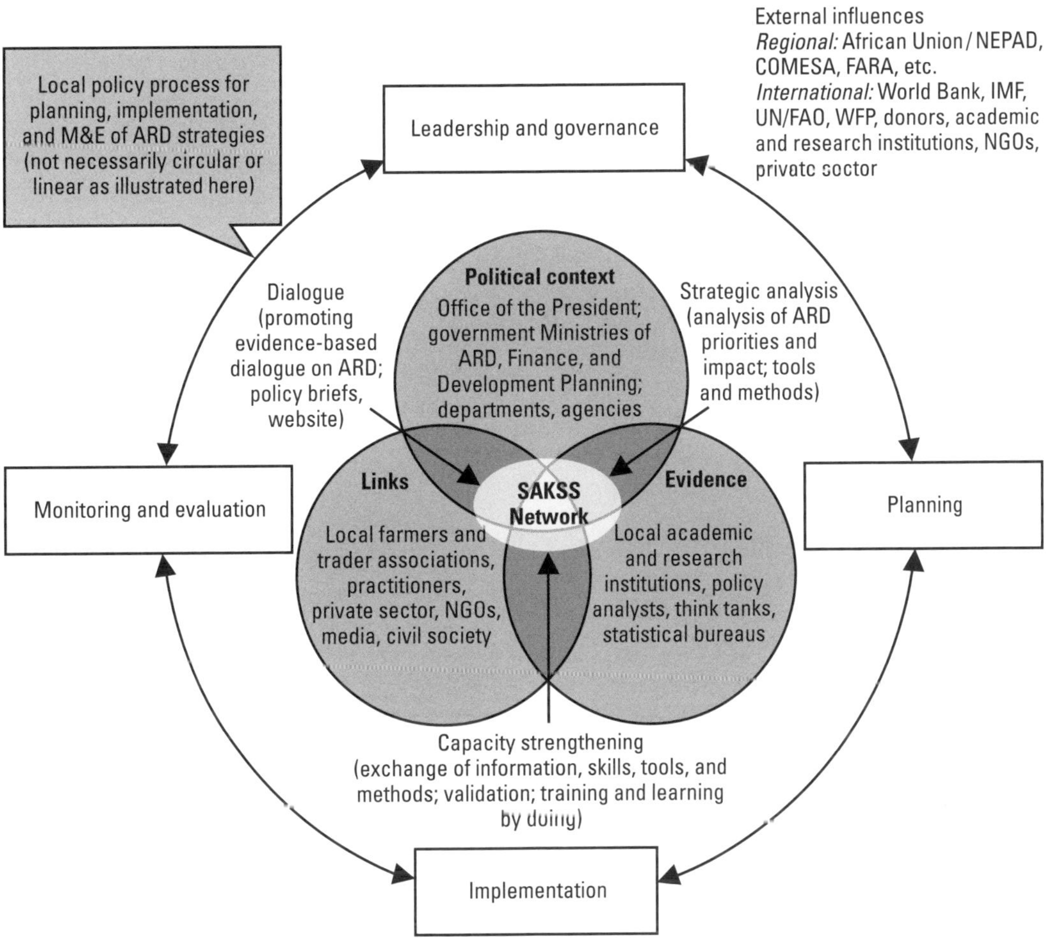

Sources: The framework is adapted from the Context Evidence and Links Framework developed by the Overseas Development Institute and described by Crewe and Young (2002) and by Young (2005).

Notes: ARD means agricultural and rural development; COMESA means Common Market for Eastern and Southern Africa; FARA means Forum for Agricultural Research in Africa; IMF means International Monetary Fund; M&E means monitoring and evaluation; NEPAD means New Partnership for Africa's Development; NGOs means nongovernmental organizations; SAKSS means Strategic Analysis and Knowledge Support System; UN/FAO means the Food and Agriculture Organization of the United Nations; WFP means World Food Programme (UN).

The greatest challenge for the leadership and governance of the ARD strategy is to improve the integration across all three activities—planning, implementation, and M&E—while allowing for sufficient flexibility to learn and improve the strategy over time. But also potentially influencing the outcome of this process is the evidence that is being generated in the center, the links that influence the national debate, and the emergence of any changing political and socioeconomic realities. In other words, as new evidence becomes available (for example, regarding lack of progress or impact), as socioeconomic and political realities change, or as new and emerging issues are brought to the forefront (for example, via civil-society groups or the media), the priorities of the ARD strategy may have to be altered.

Among the inner circles of Figure 3.2 and drawing directly from the ODI framework, the political context is the most critical because it describes the environment and process under which policies are made and thus greatly affects how evidence plays a role in it. This includes factors such as political culture, extent of civil and political freedoms, vested interests, capacities of the government to respond, and attitudes and incentives among officials (Young 2005). The overlap of political context with evidence describes the process of strategic analysis, drawing ideas and information from both government and the research community, past research, and the experience and knowledge of the actors involved. How much more closely these two will become linked also depends on how well they both overlap with where "the rubber meets the road"—the beneficiaries and interest groups of ARD strategies, such as development practitioners, farmers and trader groups, the media, and civil society in general. First, the intersection between evidence and links can be viewed as one of discourse and dialogue (for example, through publications, seminars, and the media). The intersection between the links and the political context, on the other hand, is more about advocacy, the world of campaigning and lobbying, local interest groups and the media, and the broader electorate or civil society. Finally, the external environment, including the overall socioeconomic environment, as well as the influence of regional and international actors (for example, donors), can be quite significant.

Effective linkages among evidence, dialogue, and policymaking are supposed to occur at the intersection of the three spheres. A number of case studies undertaken by ODI have generally supported this, finding that whenever there have been effective links among all three, evidence has provided a practical solution to a known policy challenge and in some cases even influenced the policy outcome (Young 2005). Other case studies have shared the same conclusion but emphasized the importance of establishing such links early in the process to ensure that the evidence is not only relevant but salient to the local context (Cash et al. 2003).

The big challenge in most African countries is to get all three to intersect given the poor state of evidence generation (due to poor-quality data, poor training and incentives, and weak peer review systems), poor links (due to poor communications and low capacities), the political context (power plays, vested interests, top-down bureaucracies, and elitist attitudes among officials), and the external environment, especially the exaggerated influence of donors (Young 2005). What the SAKSS knowledge support system strives to target and strengthen is this network of key local actors (individuals and organizations) that intersect from all three spheres. Through extensive consultations and interactions in the network, activities involving strategic analysis, capacity strengthening, and dialogue are laid out. The scope of work under each of these activities draws on the active input of all stakeholders: local research partners and analysts (evidence), key government actors and agencies (the political context), and stakeholder groups (links). With this goal in mind, and based on the practical lessons and theory discussed on bridging the research and policy divide, a number of principles can help drive this process. We review these next.

Underlying Principles and Approaches

Within the framework of knowledge support systems just described, several practical principles have been distilled from both literature surveys and experiences to date, especially in terms of their usefulness for supporting ARD strategy. These principles in turn serve as useful guidelines for establishing a new country SAKSS program.

Participation and collaboration

From the beginning, the setup of a country SAKSS program should be country-owned and driven and its processes participatory and transparent. Consequently, a continuous consultative process with key partners on both the supply and the demand side is needed. This process must include organizing regular dialogue and workshops to bring together research and analytical teams, various stakeholders, and national policymakers to share information and ideas; agree on priorities, roles, and responsibilities; and plan the activities of the program. The strategic analysis and any other knowledge management activities should be undertaken in a collaborative manner to promote local involvement and ownership. This collaborative approach demands that researchers share their choice of methodology and preliminary results at various stages throughout the process of undertaking the analysis. Only by facilitating such a continuous iterative process will the analysis and information generated remain accountable to the perspectives, concerns, and issues of both researchers as "suppliers" and policymakers and their stakeholders as "users" (Cash et al. 2003). The involvement of stakeholders early in the research design process has been shown to increase the chances of policy impact (Court and Young 2003; Ryan and Garrett 2003; Wangwe 2005).

Flexibility

Given the general consensus that there is no one-size-fits-all development model, SAKSS should always remain sufficiently flexible to adapt to different country conditions involving institutional capacity and political context, especially related to the ongoing process of strategy design and implementation. Any research and analysis activities of SAKSS must be able to take into account the broader context underlying a country's own unique development challenges and opportunities, such as natural resources and geography, stage of development and human welfare, underlying social and political conditions, and overall structure of the economy. These underlying conditions can vary widely across countries, even among neighboring countries. As described in Chapter 5, SAKSS programs have been set up in countries with different initial conditions and expectations. This has resulted in a varying sequence of approaches to analyses and knowledge systems according to each country's own unique demands and priorities. The structures of the programs and networks being established have also varied depending on the existing stock of institutional capacity and knowledge, the political context of government and donor relations, the level and source of funding, and awareness of the value of scientific evidence for policy.

High-level dialogue

Evidence-based policy research is most successfully translated into policies when the demand for policy information comes from the policymakers themselves. Policymakers pay more attention to policy research results when they are jointly generated with their own staff and when they are convinced that their interests have been acknowledged by the researchers. Implementation of policy advice is enhanced when attention is paid to continuous policy dialogue with key policymakers, executive government officials, and parliamentarians. This dialogue offers the best way to affect or influence specific policies or decisions even though it is often difficult to systematically trace how people have used the information that was generated and disseminated. The degree of proximity of outside research institutions to in-house policy analysis units (for example, within legislative and executive branches of government) has been found to have an important effect on how well research results are communicated and received by policymakers (Ryan 1999). The existence of close personal links between individuals (researchers and policymakers) can also be effective (Timmer 1998; Court and Young 2003). For example, in South Africa there has been a sizable overlap among policymakers and outside researchers, resulting in regular contact between the groups and contributing to the effectiveness of linking research results with policy dialogue (Court and Young 2003; Bhorat 2007). A potential disadvantage is overly close relationships, which can marginalize the contributions of other researchers and research institutions, limiting the diversity of views to which policymakers have access (Stone, Maxwell, and Keating 2001).

SAKSS country programs share in this experience, as described in Chapter 5. Many of the programs that have witnessed greater success in translating results into national dialogue processes have been those that have established strong links with key policymaking figures and/or government agencies that have a high degree of proximity to such figures. However, these programs have maintained a careful balance to ensure that research integrity was not sacrificed for political influence. To balance this relationship, the programs have established formal networks that encourage broad participation and transparency. The programs have been able to maintain a high degree of credibility and legitimacy by setting up a steering committee or national advisory group that represents a broad range of key stakeholders within the policymaking environment.

Credibility and legitimacy

Knowledge systems and formal knowledge networks should be structured in a way that adheres to the same criteria for credibility and legitimacy that are applied to policy research (Cash et al. 2003). Knowledge networks are credible when the participants represent shared and common institutional mandates rather than personal research interests. The degree of credibility is only enhanced when membership is limited to those institutions with a strong reputation for their expertise and for their capacity to influence the policy process (Ryan and Garrett 2003).

When the network has a broader concentration, such as rural development, the issue of legitimacy also becomes increasingly challenging. Incorporating researchers from multiple disciplines rather than solely economics is one way of increasing the legitimacy of the network. Another way is to establish closer links between foreign and local research institutions. Local institutions have the respect of the domestic policymaking community, can offer more local knowledge, are more cognizant to domestic policy concerns, and may also be viewed as less ideologically driven than foreign institutions (Jayne et al. 1999; Wangwe 2005). On the other hand, foreign institutions can enhance the credibility of the research, providing better access to international research resources and standards as well as on-the-job learning to strengthen domestic research capacity (Jayne et al. 1999). If sustained over the long term, together with sufficient higher-degree training, such efforts can go a long way in promoting and sustaining a think tank culture that effectively influences national dialogue and decisionmaking about future policies and strategies.

Sustainability

Policy research and analysis capacity have to be built incrementally and sustainably, which means that there must be ongoing support for key government policy agencies as well as encouragement of a think tank culture for producing high-quality, policy-relevant research products. Therefore, countries ultimately need to assume ownership of SAKSS from the beginning to enable its principals and tools to become institutionalized within local government agencies and research institutions over time. To accomplish this, both in-country researchers (as suppliers) and stakeholders (as end users) need to be actively engaged early on to commit to a long-term institutionalization process that involves knowledge synthesis and generation, compiling lessons from learning by doing, institutional arrangements or platforms for linking research to policy, and human and institutional capacity strengthening. As Cash et al. (2003, 8090) point out, "Strategies to promote such systems require a sufficiently long-term perspective that takes account of the generally slow impact of ideas on practice, the need to learn from field experience, and the time scales involved in enhancing human and institutional capital necessary for doing all these things." Indeed, the knowledge systems that have been the most successful in sustaining themselves over time started small. It is therefore essential that the system initially provide information around a narrow set of issues, especially those that are important and timely within each region and are consistent with development goals. Systems have also been more successful whenever sufficient attention has been given to strengthening the capacity of both suppliers and users to comprehend and value research-based evidence and policy analysis.

Capacity strengthening

The SAKSS concept is founded on the recognition that many developing countries lack the capacity to generate the reliable research-based information and analysis needed to inform and guide development strategies.[25] Therefore, strengthening the capacity of countries to provide much-needed credible information and knowledge systems for strategy development and implementation must be integral to the ongoing activities of a country SAKSS program. Without it, as pointed out earlier, the efforts of the program cannot be sustained beyond the life of the project. The core assumption is that as relevant and timely information is increasingly provided from local sources to those engaged in the policy dialogue and the design of strategies in each region, a greater appreciation of and reliance on empirical evidence will emerge and lead to sustained improvements in sector governance and policy impact over time. A SAKSS program, therefore, must play a catalytic role in developing a capacity-strengthening strategy that promotes and improves the capacities of local partner institutions best placed to apply the SAKSS concept and those of policymakers to use the results during dialogue and deliberations about future development priorities.

At least two stages are required to develop a capacity-strengthening strategy. The first stage considers both supply- and demand-side perspectives. On the *supply side,* this stage includes:

- Initially assessing the distribution of research capacities, skills, and knowledge across a variety of local institutions and organizations in each region
- Identifying the partner institutions and individuals best placed to collaborate with on SAKSS, in close consultation with country and regional stakeholder organizations
- Identifying potential areas for collaboration on SAKSS across partner institutions based on existing capacities and knowledge (data, analysis, and knowledge management)
- Outsourcing training needs to local academic institutions that can be quickly familiarized with sophisticated analytical tools to help fill the demand for such skills among a pool of local students in higher education and practitioners

On the *demand side,* this stage includes:

- Assessing and identifying specific capacity needs of partner institutions and individuals (including academic institutions) with which a SAKSS program would be collaborating in order to fill gaps, which will also help define the kinds of training, materials, and courses needed
- Building awareness and capacity more generally among policymakers and policy analysts at the country and regional levels to better comprehend the results of research and data analysis

In the second stage, training materials and courses will need to be developed to help transfer the applied tools and skills among local partner institutions and individuals on the basis of need. Capacity strengthening should initially focus on existing tools and/or components of analysis being applied by a SAKSS program, such as economywide modeling, spatial characterization of the development domains, investment and policy analysis, and impact assessment. It is the partner institutions that will take the lead over time in providing knowledge and technical know-how. As noted earlier, the involvement of local universities in training is essential to improve and expand the pool of local experts in the long run.

In order to enhance and sustain the core capacities at all levels of operation among local partner institutions or organizations, as a general principle a three-pronged approach must be used:

- *Individual-level capacities* can be enhanced by concentrating on increasing the skills of individuals to better manage the tasks at hand.
- *Organizational-level capacities* can be enhanced by helping to promote improvement of the resources and working environment of the organization to ensure that the individuals trained are capable of implementing their task.

[25] We acknowledge earlier input on this topic provided by Suresh Babu of IFPRI. See also Babu et al. (2004).

- *Institutional capacity* can be enhanced by helping to promote a broad set of improved rules, governance structures, and policies that could otherwise hinder the effective use of the acquired capacity.

CONCLUSION

Based on the previous broad literature survey and discussion of the underlying theory from which the knowledge support system concept and approach of SAKSS are derived, we offer the conclusion that its success will ultimately depend on how well it is able to establish strong ties across a diverse group of actors in the network—throughout the process of generating credible evidence, sharing the evidence, and promoting dialogue around the evidence. Its success will also depend on how well it is able to understand the political context and overall external influences unique to the host country and how it is eventually perceived by its network of local partners based on the type of activities it supports and the evidence it generates. The next chapter builds on some of these broad principles and lessons to provide a practical guideline as to how to set up a country SAKSS.

Part II: From Concept to Application

Chapter 4: A Guide to Establishing a Country SAKSS

The successful application of SAKSS depends on how well it adheres to several underlying principles and meets certain criteria, many of which were reviewed in the previous chapter:

- *There is a clear and stated demand.* First and foremost, a SAKSS must be established only when there has been a clear and stated demand for it. It is only under these circumstances that we offer these guidelines for how to go about establishing a system.
- *Local partners shape its relevance.* Key partner organizations (for example, research institutions, government ministries, universities, and NGOs) must perceive and be engaged to help fashion its relevance and utility. Only through such levels of institutional engagement will SAKSS be able to provide improved and commonly accepted approaches that can foster, enhance, and improve synergies among the varied and multiple development efforts.
- *It adapts to local conditions.* The system must be able to be institutionalized and maintained in ways that enable it to adapt to local conditions and serve as a national and regional public resource.
- *It includes broad representation of stakeholders.* Its organizational and governance structure must be established in a way that allows broad representation of key stakeholders (government entities, universities, think tanks, development practitioners, civil society, and development partners) to maintain its relevance.
- *It has strong links with a local partner(s).* The system must be able to develop strong links with local partner institutions and organizations to help strengthen their capacities to provide and sustain the SAKSS in the long run.
- *There are sufficient resources.* Finally, but not least important, it must be able to mobilize sufficient resources and interest among government and private donors to sustain it. At a bare minimum, the establishment of a country SAKSS will cost about US$300–US$500 thousand. The cost of a more comprehensive system that requires filling knowledge gaps regarding key research questions can easily rise to US$1 million per year if a number of international organizations and researchers are also involved.

With these important considerations in mind, we now review in more detail the operational steps required to set up a country SAKSS. Essentially the structure and design of a SAKSS are inherently dependent on many country-specific factors, which may include some or all of the following:

- The types of demands and expectations articulated by national stakeholders and development partners
- The local conditions of existing capacities for policy research and analysis (both human and institutional), as well as their use
- The current state of affairs with regard to policy- and strategy-making processes (for example, PRSP, approved agricultural strategy)
- The actors and organizations involved on both the supply and the demand side, especially what they perceive as the value addition of setting up a SAKSS in their own country vis-à-vis other established outfits (for example, a country may already have a well-regarded and active policy analysis group for ARD; a SAKSS would want to be linked to or even integrated with such a group)
- The existing individual, organizational, and institutional relationships among suppliers and users of policy research and information (including government and donor relations and the type and degree of private-sector and civil-society participation)
- The state of development and political commitments for improving ARD

Despite these differences, certain common steps and criteria have been found to be useful when establishing a country SAKSS (as will also become evident when we discuss Step 5). The steps are not strictly linear and might overlap or be carried out simultaneously or in a different order depending on the particular country or regional circumstances and strategy processes. The point here is to stress the critical steps to consider.

Step 1: Conducting Stakeholder Analysis and Needs Assessment

First, a preliminary survey of the current state of existing knowledge systems is needed to avoid duplication of effort and to situate the planned SAKSS in its optimal location to fill any knowledge and capacity gaps. Stakeholder analysis can be useful at this point to determine the key actors and institutions on both the supply and the demand side of policy analysis. See Start and Hovland (2004) for guidelines and World Bank (2007a) for a review of practical tools.

On the *demand side,* this would include analysis of the following:

- *Needs*—A needs assessment among key stakeholders in establishing a SAKSS (for example, an assessment of knowledge and capacity needs, institutional relationships, donor and government relations and expectations, and funding levels)[26]
- *The strategy design process*—Reviewing the status of strategy processes in the country and the region, especially with regard to their degree of reliance on evidence and analysis
- *Local policy processes and the decisionmaking environment*—A mapping of the local policy processes and decisionmaking: who makes decisions, who has influence in shaping the policies, when can evidence come to bear during the policy process, who the key players are, what the key entry points are (influence-mapping tools can be particularly useful for undertaking such assessments; for example, see Schiffer 2008)

On the *supply side,* this would essentially involve undertaking some broad surveys of the following:

- *Data availability and coverage*—Assessing local data collection efforts and their adequacy from the standpoint of national coverage, sufficient coverage of key regions or production systems in major agroecological zones, sampling methods and accuracy, and management systems, highlighting the extent to which certain data needs critical to SAKSS are unmet[27]
- *Potential local partner(s)*—Identifying local partner agencies or research institutions that are best placed (by virtue of staff expertise, location, or data collection systems in place, for example) to support the analytical and M&E agenda of a SAKSS
- *The current state of knowledge*—Reviewing available studies (completed or ongoing) by national agencies, research institutes or researchers, other donor agencies, and international agricultural research centers that will help inform any SAKSS analysis and knowledge base

[26] See Tadesse (2006) for an example of a survey of knowledge systems in Ethiopia.

[27] Examples of such a stock-taking exercises can be provided upon request.

Finally, it is critical to examine the kinds of relationships and linkages that currently exist between the demand and the supply side, such as:

- The structures of existing relationships between suppliers of policy research and information and policymaking bodies (both within government and among private, academic, and government sources)
- Linkages between different stakeholder groups (government, donor, and private-sector groups)

Network-mapping tools exist with which to undertake such assessments (for example, see World Bank 2007a and the Net-Map tool by Schiffer 2008). Although these kinds of assessments and stock-taking exercises are essential at the beginning of the process, they are rarely carried out in a comprehensive fashion, as will be seen when we discuss Step 5. The cost of a SAKSS and the demand to set one up quickly have often limited the degree to which these assessments have been undertaken. However, the benefits of these studies to a SAKSS are expected to be large enough to warrant serious attention to completing them early in the setup process. They can still have value even after a SAKSS is set up by helping to further improve its links and activities based on such assessments. Underlying the need for these studies is the likelihood that SAKSS will have a high degree of relevance and salience when it adequately addresses local needs and expectations and will thus ultimately affect policy decisions and promote greater evidence-based decisionmaking in the long run.

Step 2: Formalizing an Operational and Governance Structure

Once the stakeholder analysis and needs assessment has been conducted, the next step is to set up a good operational and governance structure for the system. The operational elements must include both an institutional and an organizational plan of action that is agreeable to the core group of stakeholders demanding the SAKSS.

Establishing the institutional structure involves deciding on a *primary host for the SAKSS.* Ideally the structure should have the following characteristics:

- Is well regarded and respected by government, academics, the private sector, civil society, and donors. This entity could have one or two members who are respected within government and nongovernmental bodies and who can serve as champions of the SAKSS.
- Has direct links with important figures or decisionmaking bodies within government. Some examples include a department within a government ministry or a semi-independent think tank.

- Has a reputation for producing quality products and promoting dialogue on key policy issues facing the country.
- Is willing to consider the SAKSS as a joint partnership effort with an internationally based implementing partner that can provide in-kind support for the day-to-day running of the system (for example, office space, administrative support, and equipment).
- Can benefit from the SAKSS by gaining skills and expertise that it would otherwise not have, ultimately being able to carry on the work of a SAKSS in the long run.
- Has a policy of open access to data and knowledge sharing.

In Mozambique, for example, partners chose for the country SAKSS to be hosted by the Economic Directorate of the Ministry of Agriculture. This selection has its pros and cons. In most African countries, the Ministry of Agriculture has a weak influence on national development strategy priorities. Therefore, placing a SAKSS within a weak ministry may have its drawbacks in terms of effectively influencing the setting of future priorities. However, by strengthening the ministry's capacity to generate evidence and clearly outline priorities needed in the sector, the system may actually improve the ministry's effectiveness in influencing decisions at a higher level. The serious lack of capacity within countries that have emerged from war, such as Mozambique and Rwanda, will probably demand this more practical approach of embedding a SAKSS within a ministry.

The overall governance of the SAKSS, on the other hand, should always be intrinsically linked to the stakeholder groups to ensure a demand-led process. Although the SAKSS may be established within a local host institution, a national advisory or steering committee composed of representatives of key stakeholder groups should determine the direction and priorities of the network. For operational purposes, a secretariat must be set up with the primary function of managing the SAKSS. The secretariat should be led by a coordinator who may report jointly to the local host institution and the Consultative Group on International Agricultural Research or other development partners under a management and administrative arrangement among the institutions. Figure 4.1 illustrates the operational and governance structure of a SAKSS, showing the relationships among the SAKSS secretariat and other SAKSS stakeholders. In the next sections we explain the role of the National Advisory Committee and the secretariat.

Figure 4.1. SAKSS operational and governance links

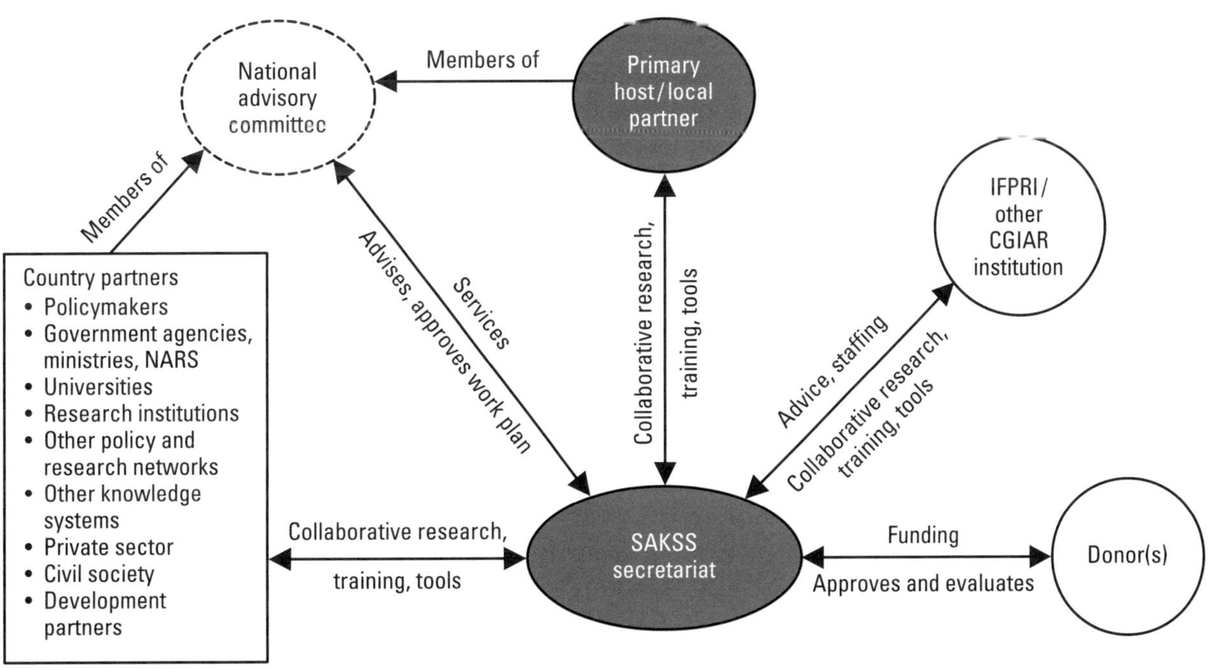

Source: Adapted from DSGD (2007a).

Notes: CGIAR means Consultative Group on International Agricultural Research; IFPRI means International Food Policy Research Institute; NARS means National Agricultural Research System; SAKSS means Strategic Analysis and Knowledge Support System.

Setting up a national advisory or steering committee

A *national advisory or steering committee* is composed of key stakeholders, including representatives from government, universities, the private sector, policy think tanks, and development partners. A broad-based committee can ensure the credibility of the system so that it can be independent and autonomous of any strong sectoral or political influence. At the same time, it ensures country ownership and relevance to local needs and expectations.

Within government, it is especially useful not to limit representation to the Ministry of Agriculture only but to include representatives from other entities, such as the Ministry of Finance. The latter not only has influence on the budget but is also just as keen to understand the role of agriculture in driving economic growth and development.

The committee would be responsible for

- establishing clear roles and responsibilities for the SAKSS and its partners,
- providing overall oversight,
- guiding the activities and operations,
- defining local needs for planning and M&E, and
- linking with broader regional and continental processes.

System management and administration

A *secretariat* composed of a coordinator and a few technical and administrative staff is often sufficient to manage a SAKSS. The SAKSS coordinator plays a critical role in running the day-to-day operations of the SAKSS. Because this individual's duties would include managing data and knowledge bases; undertaking some data analysis; facilitating a network of data providers, researchers, and analysts; and maintaining strong links with policy advisers and decisionmakers, he or she must be familiar with analytical tools and research methods in addition to having experience with knowledge management and policy processes. These skills require unique individuals who have either sat on both sides of the research–policy divide during their career or have been exposed to working with SAKSS-like entities (for example, other organizations, institutions, or networks within government or the private sector that straddle the gap between research and policymaking). Generic terms of reference for the kinds of duties, experience, and skills required of a country SAKSS coordinator or manager are provided in Appendix B.

A SAKSS work plan is typically organized around three areas:

- Strategic analysis
- Capacity strengthening
- Dialogue

These three elements together, especially once they are well integrated with local policy and strategy design processes, embody a knowledge support system as described in Chapter 3. Table 4.1 provides examples of activities that may be undertaken by a SAKSS in these three areas. These activities should be designed to be complementary. For example, any analysis undertaken should be collaborative in nature, working with local institutions and individuals in order to transfer skills and expertise in the process. The analysis activities should also take into consideration the need for dialogue to ensure timeliness and relevance. Outreach to policymakers and stakeholders before undertaking research can lead to more useful analysis and more open receptiveness to results.

In developing a SAKSS work plan, therefore, it is important that the activities be directly linked to those identified and agreed to by all key stakeholders. Final approval of a work plan should be given by the advisory or steering committee.

Step 3: Developing a Collaborative Strategic Analysis Agenda

It is critical that the process leading to an agreed set of priority areas for strategic analysis be collaborative and be negotiated between the demand and the supply side. However, there are two good reasons to maintain limited control on the supply side. First, it avoids allowing the program to become a consultancy service for one or two powerful stakeholders (for example, a government agency or single donor). Second, it preserves a sufficient level of independence to avoid political capture while maintaining scientific credibility. A strong and well-regarded local or international institution is desirable as a host for the SAKSS program because it can uphold credibility on the supply side and legitimacy on the demand side.

For international institutions, establishing legitimate ties with the demand side requires more work. It is best accomplished by forming close collaborative ties with in-country institutions or government agencies, including individuals, that are well connected to the strategy development and implementation processes in the country. For example, working with the Ministry of Agriculture and/or Finance is critical, because these ministries are often responsible for developing a country's ARD strategy. Collaboration with local technical and research institutions is desirable when the issue at hand is best handled by such institutions, including policy analysis units within government, the national statistical bureau, national agricultural research and/or policy institutes, university departments of economics or agricultural economics, independent policy research institutes or think tanks, and individual consultants.

Maintaining credibility on the supply side cannot be taken for granted. The program must never lose sight of

Table 4.1. Example of a logical framework (logframe) matrix for a SAKSS (Strategic Analysis and Knowledge Support System)

Goal	Outcomes	Performance indicators[a]	Assumptions
Stronger development performance, including increased growth, reduced poverty, improved food security, and more sustainable use of natural resources due to improved development strategies in the country	Increased investment in the agricultural and rural sector Increased agricultural productivity, particularly of low-income smallholder farmers Pro-poor rural growth and reduced rural poverty	Increased public and private investment in agriculture Increased agricultural productivity Increased smallholder commercialization Net increase in agricultural trade Export portfolio diversification Increased household income Decreased absolute and relative poverty Improved nutrition	There is a strong political commitment to the adopted strategies. The political climate allows the participation of a broad range of stakeholders. The government is willing to allocate sufficient resources to the agricultural and rural development (ARD) sector.
Development of an effective knowledge support system that builds on and integrates existing data and information, analytical tools, knowledge, and national and international analytical capacity to serve the evidence needs of national ARD-related strategy formulation and implementation processes	Enhanced knowledge, information, data, and tools for strategic analysis that informs ARD-related strategy formulation and implementation processes	Completion of analyses that address strategic issues within ARD sectors	There is low staff turnover and stability in collaborative relationships.
		Building of datasets that support strategic analyses	Sufficient, reliable national data are available.
		Development of analytical tools that enable strategic analyses and monitoring and evaluation	There are no institutional barriers or other forms of resistance to change.
	Strengthened capacity for government agencies, research institutions, and other stakeholders to supply and demand strategic analysis that informs ARD-related strategy processes Improved dialogue and communications linkages among policymakers, policy analysts, and policy beneficiaries on ARD strategy issues	Training of individuals Development of new curriculum Strategic analyses that contribute to the analytical capacity of researchers in key stakeholder institutions Evidence that ARD strategies are demanded Participation of a broader range of stakeholders in ARD sector conferences and workshops Institutionalization of communication networks among stakeholder groups Formation of technical committees and user groups Public availability and dissemination of datasets from various stakeholders Strengthening of libraries Publication and dissemination of reports and briefs	There is no political interference in data collection, use, and dissemination, especially in relation to sensitive regional economic and social information.

(continued)

Table 4.1. Continued

Goal	Outcomes	Performance indicators[a]	Assumptions
		Improved communication products that are high quality, timely, and relevant Dialogue on agricultural issues that is more evidence based	

Activities	Inputs
I. Establishment of the SAKSS	
Recruitment of coordinator and secretariat staff	Personnel
Holding an inception workshop to introduce SAKSS objectives to senior policymakers and broad stakeholders	Workshop costs NAC travel and per diem
Establishment of modalities for a National Advisory Committee (NAC)	Office overheads Rent Facilities (equipment, supplies, telecommunications, etc.)
First meeting of NAC and setting of SAKSS priorities	
Choice of host institution	
II. Strategic analysis	
Assessment of gaps in data, tools, and knowledge for conducting strategic analyses	Personnel Collaborator contracts
Assessment of evidence gaps in ARD strategy processes	In-country travel
Compilation of key ARD (and often spatially referenced) data at the micro, meso, and macro levels	Information and communication technology (ICT) platform for sharing tools and data
Compilation and refinement of practical analytical tools relevant to national planning needs	
Strategic analyses carried out in priority areas	
Development of monitoring and evaluation systems to assess country performance against planned goals	
III. Capacity strengthening	
Assessment of the capacity-strengthening needs of stakeholders who supply and demand evidence in ARD strategy processes	Personnel Collaborator contracts
Capacity strengthening of the local research community through training and collaborative work	National, regional, and international travel Training module/workshop costs
Design and implementation of training modules on key strategic analysis tools and methods	Training material Software
Institutionalization of strategic analysis tools and methodologies in national institutions (for example, government ministries and universities) through appropriate training and transfer of data and software	Equipment rental
Organization of study tours abroad for researchers and policymakers on relevant topics	
Establishment of links to the broader regional and international research community	
Support for regional learning opportunities through Regional Strategic Analysis and Knowledge Support Systems (ReSAKSSs)	

Table 4.1. Continued

Activities	Inputs
IV. Dialogue	
Assessment of stakeholders, networks, existing knowledge systems, and the general level of dialogue in ARD strategy processes	Personnel Collaborator contracts
Establishment of data-access mechanism for all researchers and analysts through a website or other data-sharing platforms	National, regional, and international travel Conference/workshop/seminar costs
Dissemination of analyses, briefs, workshop proceedings, and other publications	Publication costs
Workshops and seminars to bring together stakeholders to facilitate and inform strategy dialogue on key issues	ICT platform setup and maintenance

Source: Adapted from DSGD (2007a).

[a]As appropriate, data for all indicators should be disaggregated by region and/or other relevant social inclusion identifiers, such as gender.

the immediate need to produce high-quality products, ones that will generate constructive policy dialogue and debate, in addition to providing practical solutions for development strategy. At the same time, however, the work should remain salient to the local knowledge base and experience. To ensure credibility and salience, it is generally good practice to maintain strong partnerships with a broad set of local institutions and agencies, as the examples under Step 5 will show.

For the strategic analysis needed to support M&E and trend and spatial analysis, access to data becomes critical. It is helpful to establish strong links with the national statistical bureau, as well as other statistical units within the Ministries of Agriculture, Finance, and others (for example, Water and Natural Resources). A SAKSS is expected to help furnish the data required not only for monitoring but for the analysis and evaluation of progress, impact, and lessons learned over time.

Step 4: Developing a Knowledge Management and Communication Strategy

Most successful systems have involved people-to-people networks rather than huge systems with extremely broad mandates, so from the start it is essential to involve committed partners and local champions in building up the knowledge system. Indeed, the people aspects of partner networks remain the most important, even with the presence of ICT systems. Over time, this approach provides a unique opportunity to increase the roles of other partners (universities, research institutions, and think tanks), not only in contributing to the knowledge system but also in helping to meet the changing information and knowledge needs of a broader range of end users.

Communication is an essential component of a SAKSS. Written communications, such as policy briefs and issues papers, should be prepared on a regular basis. In addition, the SAKSS coordinator should participate in and contribute to relevant policy forums. In countries where a SAKSS has been established, the coordinators and SAKSS contributors have participated in processes relating to agricultural sector strategies and PRSPs. The coordinator must determine the best methods of dissemination, depending on such factors as type of audience, timeliness, and ICT capabilities in the country. A regionally based SAKSS can also provide support with communication processes through the sharing of regional experiences and lessons across a number of existing country-based systems.

Studies have shown that the most successful knowledge systems—linking researchers, products, and policymakers—have been those that invested significant resources in communication, translation, and mediation among researchers and end users at all levels (Creech and Willard 2001; Cash et al. 2003). Clearly, having a good communication strategy as part of the knowledge system, especially for communication with the target policymaking audience, can help foster a process of dialogue, feedback, and learning that informs the analysis and improves the effectiveness of research in policy formulation and implementation processes (Court and Young 2003).

Certain key steps are recommended to prepare and manage a good communication strategy:[28]

- stating the goal,
- analyzing the current situation and environment,

[28] From a communication strategy prepared by Marcia MacNeil, communications specialist, Development Strategy and Governance Division, International Food Policy Research Institute.

- identifying and researching the audience,
- creating messages,
- selecting appropriate vehicles for delivering the messages,
- anticipating barriers to the successful delivery of the message, and
- implementing the strategy.

It is also important to evaluate the steps, make adjustments, and ensure that the strategy is working. A generic example of a communication strategy for a SAKSS program has been provided in Appendix C.

Step 5: Developing a Capacity-Strengthening Strategy

Improvement of local capacity is critical for sustaining a credible supply of knowledge, information, and analysis over time while also generating demand by enabling a thorough understanding of its value and utility. A key function of SAKSS is to determine the capacity gaps for supplying and demanding strategic analysis and the best methods for promoting and transferring skills. The methods for transferring skills could be formal, through training workshops and support for degree programs, or informal, through collaborative research. In

Figure 4.2. Example of a SAKSS logical results framework

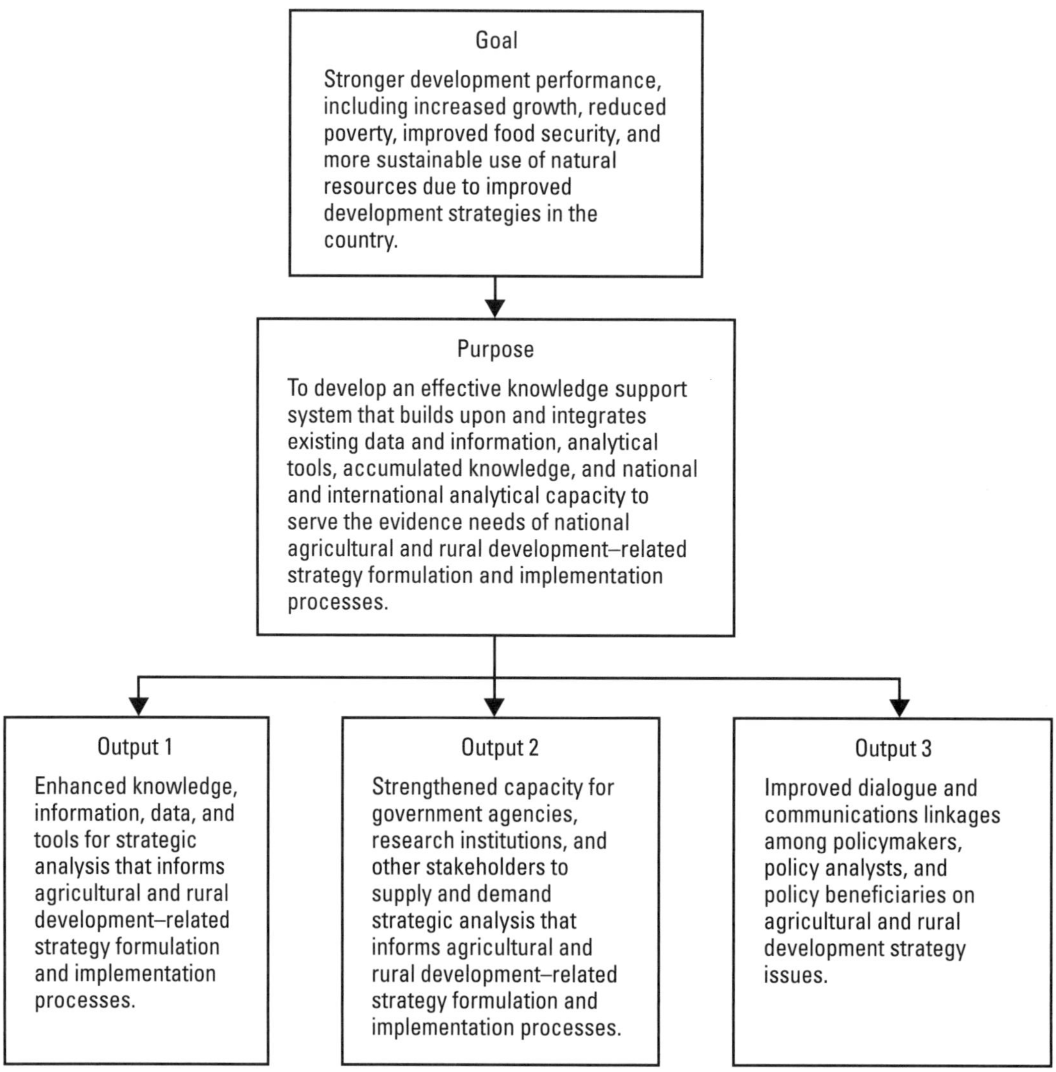

Source: Adapted from DSGD (2007a).

some cases, capacity strengthening may be needed before it is possible to proceed with analysis. For example, in Ghana a SAM was required before system designers were able to create a CGE model that would improve understanding of the dynamics of the economy. Training workshops were conducted with Ghana Statistical Services and other institutions on how to build a SAM, and CGE modeling workshops are planned for the future. An evaluation of capacity on both the supply side and the demand side would be useful before beginning any research so that capacity-strengthening activities can be well integrated with research activities. The evaluation can also serve as a baseline for monitoring the impact of the program.

Step 6: Managing for Results and Drawing Lessons

The overarching goal of SAKSS is to achieve economic growth and poverty reduction through improved development strategies. In the long run, SAKSS should contribute to increased investment in the agricultural and rural sector; increased agricultural productivity, particularly that of low-income, smallholder farmers; and pro-poor rural growth and reduced rural poverty. However, it would be difficult to quantify the effect of SAKSS on these outcomes because it would be one of a number of factors influencing strategy development processes. Moreover, exact measurements of how evidence affects strategy and policy decisionmaking are not possible. Nevertheless, SAKSS can help enable an environment that is more supportive of such results by improving the timeliness and relevance of evidence within ARD strategy processes. Monitoring and evaluation of a SAKSS should be focused on more intermediate outputs: (1) enhanced knowledge, information, data, and tools for strategic analysis; (2) strengthened capacity for government agencies, research institutions, and other stakeholders to supply and demand strategic analysis; and (3) improved dialogue and communications linkages among policymakers, policy analysts, and policy beneficiaries. These outputs are measurable through a variety of indicators and are assumed to contribute to the overall goal of improved development strategies.

Figure 4.2 provides an example of a results framework and possible work breakdown structure for a SAKSS, detailing the system's goals and outcomes. Further details of indicators that can guide the M&E of these outputs can be found in Table 4.1, which provides a sample logframe for a SAKSS. The logframe also lists potential activities that the SAKSS might undertake or coordinate. Initial SAKSS activities usually include surveys of data, knowledge, capacity, and communication gaps. These baseline studies can then serve as useful input into subsequent M&E of the impact of the SAKSS. Independent qualitative assessments of achievements and impact every few years are encouraged to capture elements that cannot be measured, such as institutional capacities, use of evidence in influencing policy dialogue, and perceptions among key target groups and stakeholders of the success and usefulness of the SAKSS. For example, SWOT analysis to identify strengths, weaknesses, opportunities, and threats can be a useful tool. Some useful and practical sources on organizational learning and the M&E of research and knowledge networks (such as SAKSS) can be found in Creech (2001) and Hovland (2003, 2007).

Conclusion

The operational steps suggested here for setting up a country SAKSS are intended to guide others wishing to do the same. However, as the next chapter will demonstrate, not every SAKSS has been set up in the same manner in practice. This is because a variety of actors (individuals and organizations) are involved in the process, with their own expectations about ownership, resources, benefits, and other issues, as reviewed in this chapter. There should always be sufficient flexibility to allow country-specific conditions to shape the need for and setup of a SAKSS. We now review the experience to date with a number of applications in Africa at both country and regional levels.

Chapter 5: Examples of Application

The following examples offer highlights of the experiences of applying the SAKSS concept at country and regional levels and describe instances in which SAKSS-based programs have provided tangible outputs that have proved timely and relevant in contributing to strategy processes (dialogue, design, and implementation). The average budget per country SAKSS, based on IFPRI's experience, has ranged between US$500,000 and US$2 million per year.

We begin with the experiences generated from IFPRI's own set of CSSPs, of which SAKSS is an integral part, and those being set up as part of the broader regional SAKSS efforts in support of CAADP implementation. The brief overview of experience covers five main areas of each program where possible:

- The process of how it was established
- Its organizational structure and donor support
- How SAKSS is integrated into the program
- Outputs generated and impact
- Future plans

After reviewing the individual programs, we provide a brief summary of any lessons and observations related to establishing a country SAKSS.

UGANDA

In Uganda there was much demand for the wealth of information and knowledge generated for almost five years under IFPRI's Strategic Criteria for Rural Investments in Productivity (SCRIP) project—the predecessor of what is now the Uganda Strategy Support Program (USSP). In recognition of the relevance of the SCRIP analyses and databases, the Ugandan government requested that IFPRI contribute to key policymaking forums. IFPRI became a member of three of the thematic working groups that reviewed and advised on development plans and M&E relating to the Plan for Modernization of Agriculture (PMA), the Ugandan government's main policy process for implementing the PRSP for the rural sector. It also began a longer-term collaboration with the Parliamentary Committee on Agriculture, a high-powered group that sets down the legislative agenda and develops the legal and regulatory frameworks and policy implementation timetable.

A number of research products came out of the SCRIP project, laying the foundation for what would become the strategic analysis component of SAKSS (for example, see Bolwig, Hazell, and Wood 2002 and the illustration in Box 1.2). A series of briefs covering development domains, commodity-related growth scenarios, and production issues in Uganda is available at the SCRIP website, in addition to various databases, reports, and presentations from the project.[29] More recent SAKSS-related research has included a study on the impact of public expenditure on agricultural growth and rural poverty (Fan, Zhang, and Rao 2004). The public investment studies and other research findings under USSP have been cited in key policy documents in Uganda, including the Poverty Eradication Action Plan and the Marketing and Agro-Processing Strategy. Moreover, USSP's research on spatial analysis in Uganda has generated a lot of demand, most recently within the framework of the development and institutionalization of a National Spatial Data Infrastructure in Uganda.

How much these efforts have contributed to national dialogue and influenced decisions is not yet clear. The initial thrust of this SAKSS-like program has mostly been a supply-driven process, providing strategic analysis of key investment options and research results defined mostly to serve donor interests and concerns. As such, the program has wavered between serving some needs in government and maintaining an overall level of irrelevance to the setting of future national priorities.

More recently, however, the program has since set up a country SAKSS within a government department, the PMA, working closely with national partners, donors, and the Ministries of Agriculture and Finance. This move may very well enable research by international institutions such as IFPRI to have a more direct impact on the national dialogue and decisionmaking about future development for ARD in Uganda. Finally, at the time of this writing, the director of PMA also happens to be a former coordinator of the ReSAKSS–East and Central Africa (ECA) node, providing a champion within the government to help steer the direction of the program while also maintaining strong ties with regional efforts to promote the sharing of experiences across countries in the region.[30]

---

29 <http://www.foodnet.cgiar.org/SCRIP/index.htm> (accessed July 28, 2009).

30 The ReSAKSS-ECA, a regional SAKSS network in East and Central Africa, is discussed in the section "Africawide efforts to support CAADP and regional strategy."

Ethiopia

In stark contrast to the situation in Uganda, the Ethiopia Strategy Support Program (ESSP) was originally developed in 2004 upon the request of the Government of Ethiopia, with leadership from the economic adviser to the prime minister, the minister of agriculture, and the state minister of finance and economic development. A NAC was formed very early in the process. The NAC has been instrumental in setting ESSP priorities for knowledge management, research, and capacity building ever since. The institutionalization of the project within a semiautonomous policy research institute and close to the prime minister's office was also a direct result of decisions led by the NAC. ESSP is directly linked with the Ethiopian Development Research Institute (EDRI) and collaborates with it on research and training workshops. The program is housed in the main IFPRI-Ethiopia office on the campus of the International Livestock Research Institute (ILRI) in Addis Ababa but maintains an active research and outreach base at EDRI. Following the organizational structure of a SAKSS as presented in Chapter 4, Figure 5.1 illustrates the operational and governance links of ESSP, including its network of partners.

The program structure is organized around three major pillars:

1. A rural economy knowledge support system (REKSS) to integrate knowledge, conduct analyses on the rural economy, and track the impact of rural investments and activities
2. Actionable applied research to improve or fill gaps in knowledge on rural development
3. Policy analysis capacity strengthening through targeted training and related activities

The overall organizational structure of ESSP emphasizes the need to preserve a degree of independence and identity as a think tank in order to maintain the role of the implementing institution, in this case IFPRI. The REKSS component, on the other hand, is intended to serve more directly as the outreach and servicing arm of ESSP, organized around the SAKSS concept. The REKSS includes a GIS laboratory housed at EDRI and carries out training workshops, seminars, and other capacity-strengthening activities through EDRI and together with activities under the other two pillars.

This CSSP model, with a REKSS or SAKSS component, has proved useful for IFPRI to adopt elsewhere in Africa because it helps maintain IFPRI's own institutional identity and reputation for independent research while ensuring the relevance of the research and activities it undertakes as part of a SAKSS. Therefore, although many of the CSSPs have subsequently been set up as IFPRI or country programs, their role as secretariats of SAKSS networks has ultimately helped maintain a degree of credibility in the outputs produced.

Results from macroeconomic modeling and simulations carried out under the ESSP served as critical inputs in the Government of Ethiopia's own efforts to re-establish the importance of staple foods for future ARD investments in the country. The modeling work advanced the analysis of progress toward the MDGs and created substantial demand for further MDG work through the ESSP by the government and donors. The first *Atlas of the Ethiopian Rural Economy* was also produced in collaboration with EDRI. An external review referred to the *Atlas* as an important research source and one that contributes well to future institutional building efforts (Colman and Mellor 2007). Another important ESSP product has been the paper "Growth Options and Poverty Reduction in Ethiopia: An Economy-wide Model Analysis" by Diao and Nin-Pratt (2007). This study examined which subsectors within agriculture had the best potential to improve growth and poverty reduction.

In response to the REKSS team's outputs, IFPRI researchers have been contacted by both national (Addis Ababa University, Electrical Power Corporation) and international (World Bank, Netherlands Development Organization, Natural Resources International Limited) policy researchers and analysts for consultations and input into their ongoing work. Also, local media have cited the team's work on enhancing access to spatial and nonspatial information on the national rural economy. The REKSS team has been asked to take a leading role in establishing a national professional society for GIS for Ethiopian development. The databases created and maintained by the team have been requested by policy analysts, researchers, and students from many regions in the country. In addition, the training materials developed for ESSP's workshop series Spatial Analysis for Rural Economic Policy have been much requested and have been the basis for several other training activities conducted independently.

The program has since attracted a new round of commitments by the Government of Ethiopia, IFPRI, and donors to extend ESSP into a five-year second phase from January 2008 to December 2012. The program's second phase is structured to increasingly expand the capacity-strengthening pillar to prepare for the eventual transition to full country ownership. During the second phase more than 60 percent of the activities of the program will be launched from the lead partner institution, EDRI, and by 2010 about 80 percent of the project activities are expected to be implemented by Ethiopian professionals.

Ghana

Building on previous IFPRI projects in Ghana and several months of preliminary missions and discussions with potential stakeholders and collaborators, the Ghana Strategy Support

Figure 5.1. Ethiopia Strategy Support Program (ESSP) operational and governance links

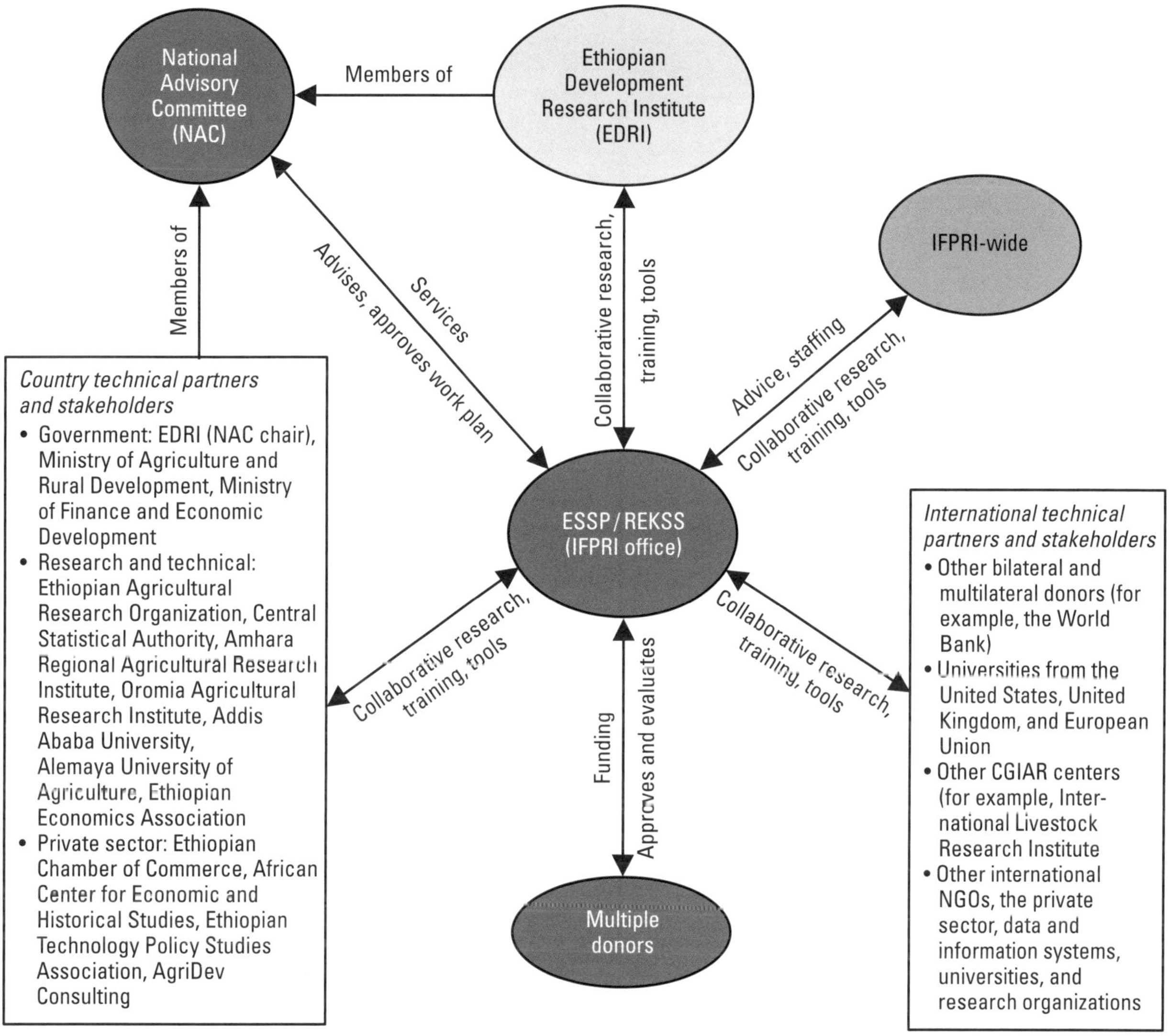

Source: DSGD (2006).

Notes: This figure is intended to reflect the more general structure presented previously in Figure 4.1. CGIAR means Consultative Group on International Agricultural Research; ESSP means Ethiopia Strategy Support Program; IFPRI means International Food Policy Research Institute; NGO means nongovernmental organization; REKSS means rural economy knowledge support system.

Figure 5.2. Ghana Strategy Support Program (GSSP) operational and governance links

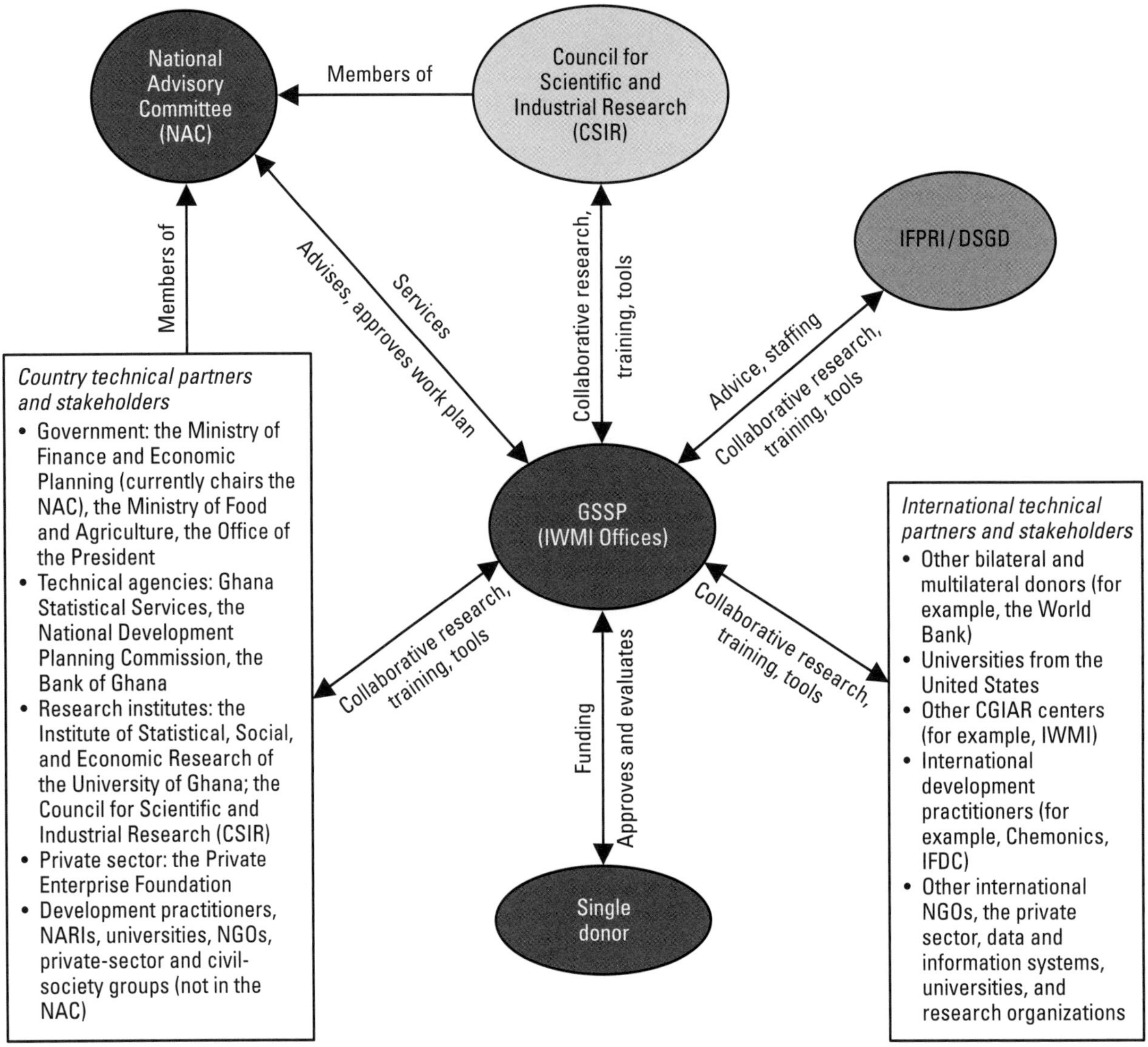

Source: DSGD (2005).

Notes: This figure is intended to reflect the more general structure presented previously in Figure 4.1. CGIAR means Consultative Group on International Agricultural Research; DSGD means Development Strategy and Governance Division; GSSP means Ghana Strategy Support Program; IFDC means International Center for Soil Fertility and Agricultural Development; IFPRI means International Food Policy Research Institute; IWMI means International Water Management Institute; NARI means National Agricultural Research Institute; NGO means nongovernmental organization.

Program (GSSP) was established in 2005 with high-level support from the Ministry of Finance and Economic Planning (MoFEP) and the Ministry of Food and Agriculture (MoFA). The Ghana program began under the same auspices as the one in Uganda; it was initially set up at the request of a single donor. But unlike the Uganda program, it was set up with the establishment of a country SAKSS in mind. The initial groundwork to set up a SAKSS involved many consultations with government entities, donors, and other stakeholders (universities, think tanks, and nongovernmental bodies and organizations). This groundwork was important in gauging whether there was sufficient demand for a SAKSS and, if so, how to go about setting up the system so that there would be sufficient country ownership. Based on these initial consultations and negotiations, it was agreed that IFPRI should establish a country support program, one that would be based on the model in Ethiopia. This type of setup would allow IFPRI to maintain an active in-country presence in order to deliver quality outputs and deliver on the expectations inherent in the SAKSS concept.

Today GSSP is considered a research, communication, and capacity-strengthening program of IFPRI in Ghana. The program has core funding from the U.S. Agency for International Development–Ghana but with a mandate to develop a multidonor-funded program. Committed to the expectations of a SAKSS-like program, GSSP is working with local stakeholders to help with generating information, improving dialogue, and sharpening decisionmaking processes essential for effective formulation and implementation of development strategies. Following the organizational structure of a SAKSS as presented in Chapter 4, Figure 5.2 illustrates the operational and governance links of GSSP, including its network of partners.

Initial research was completed on development opportunities in Ghana using spatial analysis and a study of growth scenarios based on an economywide multimarket model. The modeling work in particular has been in great demand by the government and was prioritized by the program's NAC, made up of representatives from government, research institutes, and the private sector. Several capacity-strengthening training sessions relating to modeling have been successfully conducted, including SAM-building and multiplier analysis workshops with the Ghana Statistical Services, MoFEP, and the Bank of Ghana among other institutions. In the second year of the program, research expanded to include studies on the returns to public investment, the effectiveness of public service provision, and options for strengthening capacity at MoFA.

The program leader of GSSP contributed to the dialogue held during the process of revising Ghana's Food and Agricultural Sector Development Policy (FASDEP). The revised FASDEP also cites one of the collaborative studies produced by the SAKSS program, "Regional Disparities in Ghana: Policy Options and Public Investment Implications" by Ramatu Al-Hassan of the University of Ghana–Legon and Xinshen Diao of IFPRI (Al-Hassan and Diao 2007). This study was presented both nationally and internationally to solicit feedback.

Nigeria

Recognizing the knowledge, capacity, and communication weaknesses of Nigeria's Federal Ministry of Agriculture and Water Resources (FMAWR), IFPRI developed a concept note suggesting the development of a SAKSS-type program in Nigeria in late 2004. After consultations with stakeholders and stock-taking activities, IFPRI launched the Agricultural Policy Support Facility (APSF) in partnership with FMAWR in March 2007 with financial support from the Canadian International Development Agency. The APSF is now an initiative of IFPRI's Nigeria Strategy Support Program. The purpose of APSF is to strengthen the capacity of the federal government and researchers in Nigeria to design and implement evidence-based, pro-poor, gender-sensitive, environmentally sustainable ARD policies and strategies. The desired outcomes include improved policy analysis for rural and agricultural development, enhanced capacity in Nigeria to undertake such analysis, and increased and improved consultations on agricultural policy issues among key stakeholders within and outside of government.

APSF engages in two major activities:

- Collaborative policy research to address specific pressing knowledge gaps or growth options for attaining development targets; options for agricultural-sector growth that is pro-poor, gender sensitive, and environmentally sustainable; and ensuring that vulnerable households benefit from agricultural and rural development and economic growth
- The establishment of a REKSS to improve the enabling environment for ARD-related strategy formulation and implementation and work toward the establishment of a Nigeria-owned REKSS

Research activities are currently under way, and preliminary findings were presented at a stakeholder workshop, Developing Evidence for Agricultural and Rural Development Policies and Strategies in Nigeria, held in May 2008 in Abuja. The workshop brought together policymakers, researchers, farmer organizations, and development partners. Several briefs on the research were distributed at the workshop, with subjects including agricultural policies and strategies, public expenditure on agriculture, and an evaluation of the impact of a community development project. A major challenge in Nigeria is access to existing information for policy design and implementation. APSF is working with information managers within FMAWR and with the librarians of lead agriculture universities and research institutions to strengthen their capacity for sharing policy research information.

The APSF program has been highly consultative in identifying the research gaps, capacity-strengthening needs, and communication linkages. In addition to three consultation workshops and two stakeholders' workshops, the collaborative research teams have held individual consultation meetings with staff of government agencies, research institutions, extension programs, development partners, and the private sector. Many of these teams have met with the same people but held very different detailed discussions. The collaborative research teams consist of IFPRI researchers and representatives from locally established institutions, mainly universities. This collaborative approach helps to strengthen the capacity of the institution or university faculty. This new capacity among faculty could be transferred to university students if the new tools and methods were incorporated into the curriculum.

Several questions have arisen from the Nigerian program experience that may be of value to consider when establishing a SAKSS: When is there too much consultation? Is there an approach that can maintain the high level of consultation but minimize the time burden of those involved, especially those being consulted individually or through workshops? Finally, is it more effective and sustainable to work specifically to strengthen the capacity of one research institution, such as the Ethiopian program, or to spread the capacity strengthening through collaborative research to a large number of institutions?

RWANDA

Rwanda, as the first country to hold a CAADP roundtable, is at the forefront of countries aligning their development strategies with the CAADP framework. The CAADP roundtable involves the review of ongoing and future national development efforts in alignment with the CAADP objectives and principles (Badiane, Odjo, and Ulimwengu 2010). As part of the CAADP implementation process, a Rwandan country SAKSS will be set up to help provide the information and knowledge support necessary to guide and review the progress of agricultural-sector performance. In preparation for the roundtable, stock-taking reports and analyses were undertaken to answer critical questions, such as "Agricultural Growth and Investment Options for Poverty Reduction in Rwanda" (Diao et al. 2007b). The country SAKSS will naturally build on this work by coordinating research, capacity strengthening, and dialogue to further address policy questions that will arise as the new economic and agricultural-sector development strategies are implemented.

Since the signing of the compact to this end in March 2007, the process of setting up a country SAKSS has moved very slowly, in part due to the demand that it be maintained as an integral part of any post-CAADP compact follow-up in the country. At the time of this writing, efforts are well under way to establish a Rwanda SAKSS and hire a coordinator following a number of consultations among Rwandan partners, donors, IFPRI, and the ReSAKSS-ECA node at ILRI in Nairobi (see "Africawide efforts to support CAADP and regional strategy").

MOZAMBIQUE

A number of Consultative Group on International Agricultural Research (CGIAR) partners, led by IFPRI, began serious efforts in 2007 to set up a country SAKSS in Mozambique following a request from the Ministry of Agriculture and the Swedish Embassy through the Swedish International Development Cooperation Agency. The CGIAR centers involved, aside from IFPRI, included the International Water Management Institute (IWMI) and the International Crops Research Institute for Semi-Arid Tropics (ICRISAT). Both centers are leading the establishment of a ReSAKSS—SA in southern Africa.[31] The Mozambique SAKSS was envisioned from the start as an effort intended to help strengthen the ability of the Ministry of Agriculture (and partner institutions) to provide credible information during the design and implementation of the country's ARD strategy. At the time, Mozambique was in the process of designing a new agricultural strategy slated to be completed by the end of the year. To be timely and useful, a set of proposed strategic analysis and research activities was planned for the first year, combining short-term and demand-based approaches for strategic analysis that will help identify key investment and policy alternatives for agriculture in Mozambique. The work plan was reviewed and validated by several local key partners.

The primary activities in the work plan included the following:

- *Monitoring and evaluation* of agricultural investments and performance
- *Strategic analysis* of future investment and policy options, including short-term studies on emerging issues
- *Knowledge sharing and policy dialogue* to further promote local capacities for generating and sharing information and analysis with policymakers to strengthen the utility of evidence-based decisionmaking

[31] ReSAKSS-SA, a regional SAKSS network in southern Africa, is discussed further in the section "Africawide efforts to support CAADP and regional strategy."

- *Capacity strengthening* (an integral part of all four activities) to strengthen local capacities for undertaking strategic analysis and knowledge support during the process of designing and implementing ARD strategies in Mozambique

In the year of its inception, the program organized a number of seminars based on preliminary results of ongoing work undertaken with local partners. The process of setting up a NAC also began, especially after the Ministry of Agriculture formerly endorsed the program and requested that the SAKSS be hosted by its economic directorate. The program has since expanded to include the appointment of an IFPRI CSSP program leader in addition to a SAKSS coordinator, both of whom will work under the overall guidance of a NAC. The program leader is essentially helping to manage the overall program and maintain research quality and oversight. A SAKSS coordinator has also been appointed to ensure that greater information flows and closer ties are established for strategic analysis, capacity strengthening, and policy dialogue among stakeholders in the SAKSS network.

Africawide Efforts to Support CAADP and Regional Strategy

At the Africawide level, three ReSAKSS nodes have been set up in each of the major regional economic communities (RECs): the Common Market for Eastern and Southern Africa, the Economic Community of West Africa States, the Southern Africa Development Community in collaboration with five CGIAR centers (IFPRI, ICRISAT, IWMI, ILRI, and the International Institute of Tropical Agriculture, IITA). The ReSAKSS was established in 2006 to provide analytical and knowledge support for the CAADP roundtable process and the establishment of a country SAKSS in the postroundtable period.

The ReSAKSS nodes are located at three Africa-based CGIAR centers: IITA in Ibadan, Nigeria; ILRI in Nairobi, Kenya; and IWMI in Pretoria, South Africa, run jointly with ICRISAT. IFPRI is helping to coordinate a common agenda across the three nodes, providing technical and analytical support and helping maintain and strengthen links with a broad network of CAADP partners. The nodes are governed via steering committees chaired by the respective RECs. The steering committees provide overall oversight and help ensure that the ReSAKSS agenda remains relevant to development priorities, CAADP, and regional strategies. Table 5.1 summarizes the network of partners of ReSAKSS at the various subregional-level nodes, including the respective steering committee chairs. Typically the Africawide and international-level network partners are linked with all three nodes.

A working group composed of the coordinators meets twice a year to develop and review in detail an integrated work plan and communication and outreach strategy of the ReSAKSS, as well as to share experiences of implementation. Each full-time coordinator oversees the day-to-day research, analysis, and outreach activities of his or her node, coordinating with other nodes under the guidance of an Africawide coordinator at IFPRI to ensure consistency, avoid duplication, and fill any gaps based on each individual node's comparative skills and existing networks. This has involved working with a diverse group of local and international partners (including other CGIAR centers) to build on existing analyses and expertise that can provide the state-of-the art strategic analysis required to promote evidence-based dialogue concerning the CAADP investment priorities.

In close collaboration with the African Union and NEPAD, RECs, and member states, ReSAKSS focuses its activities in the following three areas:

- Strategic analysis
- Knowledge management and communications
- Capacity strengthening.

The strategic analysis activities help fill critical knowledge gaps in assessing progress toward realizing the CAADP goals of allocating 10 percent of the national budget to agriculture, achieving a 6 percent annual agricultural growth rate, and meeting the first MDG of halving poverty and hunger by 2015. As part of this effort, IFPRI has been assisting ReSAKSS with developing an M&E framework to inform CAADP implementation (Benin, Johnson, and Omilola 2010). The theoretical framework is presented in Figure 2.2.

In the knowledge management and communications component, ReSAKSS and its network of partners collect data on key indicators such as public spending; integrating and building upon existing data, analytical tools, and knowledge; and facilitating timely access to the knowledge by African policymakers and development partners to allow for more evidence-based decisionmaking. To this end, ReSAKSS has launched a website to share the information it generates and compiles on these key indicators and on ARD in general (see Figure 3.1). The site also provides the latest updates on CAADP implementation, tools, and key events related to ARD priorities and M&E.

Finally, ReSAKSS is also helping to build and strengthen institutional and technical capacity by promoting collaboration in generating and disseminating data and providing access to knowledge and information products. A key element under capacity strengthening has been to provide technical support to the CAADP roundtable process and to the setup of country SAKSSs at countries' requests. Technical support has involved

Table 5.1. Key network partners of the ReSAKSS

Level of partnership	Africawide	East and Central Africa (ECA)	Southern Africa (SA)	West Africa (WA)
Steering committee chair	n.a.	Common Market of East and Southern Africa (COMESA)	Southern African Development Community (SADC)	Economic Community of West African States (ECOWAS)
Regional	African Union's New Partnership for Africa's Development (NEPAD) secretariat; COMESA; ECOWAS; SADC; Forum for Agricultural Research in Africa; Association for Strengthening Agricultural Research in Eastern and Central Africa (ASARECA); West and Central African Council for Agricultural Research and Development (CORAF/WECARD); African Development Bank	East African Business Council; Eastern African Farmers Federation; ASARECA; COMESA business council; East African Community secretariat; country representatives (ministries of agriculture, universities, National Agricultural Research Institutes [NARIs], nongovernmental organizations [NGOs], private-sector groups)	COMESA; Southern African Confederation of Agricultural Unions; Regional Universities Forum for Capacity Building in Agriculture; Food, Agriculture, and Natural Resources Policy Analysis Network; University of Pretoria, South Africa; country representatives from Malawi, Mozambique, and Zambia (ministries of agriculture, universities, national statistical bureaus, NARIs, NGOs, private-sector and trade associations)	West African Economic and Monetary Union; Comite Permanent Inter-etats de Lutte Contre la Secheresse dans le Sahel; CORAF/WECARD; Conférence des Ministres de l'Agriculture de l'Afrique de l'Ouest et du Centre (CMA/AOC); Réseau des Organisations Paysannes et de Producteurs de l'Afrique (a farmers' group); country representatives (ministries of agriculture, universities [for example, the University of Ibadan], NARIs, NGOS, private-sector groups)
International	Consultative Group on International Agricultural Research centers (the International Water Management Institute, International Institute of Tropical Agriculture, International Livestock Research Institute, and others) Universities from the United States, the United Kingdom, and the EU United Nations bodies (the Economic Commission for Africa and the Food and Agriculture Organization) and development partners (the World Bank, Global Donor Platform for Rural Development, U.S. Agency for International Development, U.K. Department for International Development, Swedish International Development Cooperation Agency) Bill and Melinda Gates Foundation Alliance for a Green Revolution in Africa Famine Early Warning Systems network			

Source: Authors.

Box 5.1: Strategic analysis to inform Comprehensive Africa Agriculture Development Programme (CAADP) roundtables

In order to determine future growth options and the level of effort required to achieve growth, poverty reduction, and food security targets, a series of logical questions was used to help guide the analysis:

- Is a 6 percent growth rate in agriculture sufficient to reach the Millennium Development Goal (MDG) for poverty reduction in the country? If not, what would be the required rate of agricultural growth (the MDG growth rate)?
- What are the most promising sectors and subsectors in which to accelerate growth in order to reach the 6 percent or required MDG growth target?
- What level of investment is necessary to accelerate growth in these sectors and subsectors to reach the 6 percent or required MDG growth rate?
- What other additional investments are needed, and in which areas, to support growth in the priority sectors and achieve the 6 percent or required MDG growth rate?
- How much of the required investment would a budget share of 10 percent be able to cover?
- What would be the funding gap to meet the required investments?
- What additional measures need to be taken to ensure that the sector with the faster growth priority translates into better distribution and nutrition outcomes?

The analysis adopted three primary methodologies to answer the questions: (1) economywide simulation analysis combined with detailed spatial disaggregation of economic subsectors and subnational regions to determine sources of growth and poverty reduction, (2) assessment of public investment options based on their relative economic rate of return, and (3) estimation of the total public resources required. The level and detail of analysis varied depending on the availability of data and already existing analytical work in these countries.

Source: DSGD (2007b).

facilitating in-country stock-taking exercises and strategic analysis to determine what it will take to achieve the 6 percent growth target and the poverty MDG by 2015, including whether the 10 percent budget allocation to agriculture is sufficient. The setting up of country SAKSS is intended to allow for follow-up to roundtable outcomes to inform subsequent policy debates and decisionmaking related to implementation.

The country-level analyses that have been undertaken to inform the country CAADP roundtables involved Rwanda in 2006; Kenya, Malawi, Uganda, and Zambia in 2007; and a number of West African countries in 2008.[32] A series of logical questions served to guide the strategic analysis, as discussed in Chapter 2, to determine future growth options and the level of effort required to achieve growth, poverty reduction, and food security targets (Box 5.1). The results of the analysis have been received well within each country, having been undertaken in collaboration with in-country partners. They are serving as critical input into the stakeholder dialogue and the preparation of a roundtable compact within each country.

Finally, efforts to set up country SAKSS as part of the CAADP agenda have been closely tied with the examples for Mozambique and Rwanda given earlier. In 2009, a similar

[32] Progress with the CAADP analysis and roundtable process can be found on the ReSAKSS website, <www.resakss.org> (accessed July 28, 2009). Among the country studies undertaken, see, for example, Breisinger et al. (2008) on Ghana; Thurlow, Kiringai, and Gautam (2007) on Kenya; Benin et al. (2008a) on Malawi; Thurlow (2008) on Mozambique; Diao et al. (2007b) on Rwanda; Thurlow et al. (2008a) on Uganda; and Thurlow et al. (2008b) on Zambia.

effort also began in Malawi. In these and other IFPRI CSSPs, strong links and information flows are being maintained between the country and regional efforts. This is providing an opportunity for cross-country learning, training, peer review of progress with CAADP implementation, and promotion of evidenced-based dialogue on emerging regional issues facing ARD (for example, regional market integration and technology spillovers).

Summary of Early Lessons Learned

The country and regional examples illustrate how the SAKSS concept has been applied differently in response to local conditions. SAKSS is meant not to replace any other existing country or regional networks but instead to strengthen and build on them. In some countries there have been stronger capacities among existing networks and institutions, providing critical data and analysis for policy dialogue and strategy formulation. Within each country, the interplay among the local political context, the individual relationships and capacities of local partners and stakeholders, the perceptions of what SAKSS is, and the perceived credibility of IFPRI and individuals introducing the system have been influential in shaping the purpose and activities of a country SAKSS.

Among the examples reviewed in this chapter, many are still at very early stages in their development. Those that have become established as an integral part of IFPRI's country support programs have tended to focus more attention on assisting local partners with improving data, spatial analysis, and development of final products (for example, the *Ethiopia Atlas* for ARD). Much of the other attention has been on generating timely and relevant research outputs (working papers and briefs) in order to enrich the dialogue on emerging ARD issues facing the country. Overall, the country SAKSSs have maintained a strong institutional identity with IFPRI, especially those that have remained an integral part of IFPRI's CSSPs. Recent exceptions have been in Uganda and Mozambique, where a SAKSS coordinator was placed within a government agency or ministry. A similar model is being considered in Malawi, where a SAKSS coordinator is situated within a key ministry working on ARD issues but maintains close ties with an IFPRI CSSP leader. In Malawi and Mozambique, the role of the IFPRI CSSP leader is primarily to guide and maintain research quality while representing IFPRI's own goal of strengthening ARD policy research capacities and impact on poverty and hunger reduction. Because this model makes sense only where there is an IFPRI CSSP, less costly models will be needed if more countries demand to set up a SAKSS as part of their commitment to CAADP. Rwanda has taken the lead in this area by placing in the Ministry of Agriculture a SAKSS coordinator who operates as part of the broader ReSAKSS support to CAADP.

Among the programs, certain early observations and experiences are worth noting, all of them based on the direct observations and experience of the authors and on input from interactions with colleagues implementing SAKSS at IFPRI and other CGIAR centers:

- Dialogue with the local policymakers, analysts, and existing networks is essential at the early stages to determine local needs and capacity.
- Having local champions of SAKSS has an important impact on the effectiveness and speed with which SAKSS and ReSAKSS nodes are set up. For example, the new director of the Ugandan PMA, who is a former ReSAKSS coordinator, has helped set up a SAKSS that is more directly linked with the secretariat.
- Stronger ties with local partner institutions and government bodies help maintain the relevance of SAKSS as country led and owned.
- The ability to maintain quality products that have been produced in close collaboration with local partners and institutions raises the credibility of the program while also maintaining its relevance among stakeholders.
- The presence of an active champion on the steering committee helps to establish stronger ties among the network data and analysis suppliers, as well as the relevance of their outputs among users (for example, government agencies, policymakers, and development partners).
- The existence of multiple donors and a sufficient level of resources strengthen the perception of SAKSS as a public good that supports local interests and capacity needs.

At the time of this writing, no comprehensive evaluation of SAKSS has taken place, aside from the evaluation of two country support programs (those in Ethiopia and Ghana). Therefore, the observations shared here remain largely unsubstantiated. A more thorough external (or independent) evaluation of the SAKSS concept and its applications is required.[33] From this practical lessons can be drawn to improve the current and the next round of SAKSS-like networks at both the country and the regional levels.

[33] The exception was an internal review of SAKSS undertaken as part of an overall evaluation of the Development Strategy and Governance Division at IFPRI in 2007. The results of the evaluation were mixed. At one end, the reviewers struggled to understand how the SAKSS concept fits within the research mandate of IFPRI. At the other end, the need for a SAKSS-like network in Africa was undeniably recognized.

Conclusion

The applications of the SAKSS concept at both the country and the regional levels show some diversity in structure and activities related to both strategic analysis and knowledge support systems. These have been affected by the process of how each SAKSS was established, the actors involved and their interrelationships, the unique policy processes for ARD strategies in each country and at the regional level, the expectations of stakeholders, and perceptions of what SAKSS is adding value to, among others. Many of the applications reviewed here have been in place for a number of years. Therefore, lessons are beginning to surface. However, there is yet to be a thorough external evaluation of the concept across all current applications—particularly as it relates to the goals of SAKSS. This can have important implications on the direction SAKSS should be taking in the future. We explore this issue a bit in the summary conclusion of this technical guide.

Chapter 6: Summary Conclusion

The SAKSS concept is intended to provide a framework that promotes linkages between analyses and decision-making and thus helps to maintain a strategic approach in the research and analysis undertaken. This technical guide has provided background on the SAKSS concept, a review of the kind of tools and approaches needed for the knowledge support system, and a guide as to how to go about setting up such systems at the country level. It began with a preface that described the evolution of the SAKSS concept from the time of its inception through its transformation into IFPRI CSSPs and regional nodes in Africa.

The application of SAKSS as a network managed by a secretariat at the country level (for example, via the CSSPs) and at the regional level (for example, via the ReSAKSS nodes) allows lessons to be drawn and thus improve our understanding of how such systems can be made more effective in helping to bring evidence to bear during policy dialogue and decisionmaking about current and future ARD strategies. As noted in Chapter 5, it is still too soon to determine the success of these systems without a more detailed comparative analysis, especially if it can be derived from an external independent evaluation. Nevertheless, in the interim, certain lessons and principles have been highlighted that can serve as a guide for implementing a successful SAKSS. We summarize some of these here:

- *Local partners must shape the relevance of a SAKSS.* Key partner organizations (for example, research institutions, government ministries, universities, and NGOs) must perceive and be engaged to help fashion its relevance and utility. Only through such levels of institutional engagement will a SAKSS be able to provide improved and commonly accepted approaches that can foster, enhance, and improve synergies among the varied and multiple development efforts. Dialogue with the local policymakers, analysts, and existing networks is essential at the early stages to determine local needs and capacity.
- *A SAKSS should adapt to local conditions.* It must be able to be institutionalized and maintained in ways that enable it to adapt to local conditions and serve as a national and regional public resource.
- *A system must maintain broad representation of stakeholders.* Its organizational and governance structure must be established in a way that allows broad representation of key stakeholders (government entities, universities, think tanks, development practitioners, civil society, farmer organizations, and development partners) to maintain its relevance.
- *A SAKSS should establish strong links with a local partner(s).* It must be able to develop strong links with local partner institutions and organizations to help strengthen their capacities to provide and sustain the SAKSS in the long run. Strong ties with local partner institutions and government bodies help maintain the relevance of a SAKSS as country led and owned.
- *A system must collaboratively produce quality products.* The ability to maintain quality products that have been produced in close collaboration with network partners and institutions increases the credibility of the system while maintaining its relevance and utility among partners and stakeholders.
- *A SAKSS needs a local champion.* The presence of an active champion on the steering committee helps to establish stronger ties among the network data and the analysis suppliers, as well as the relevance of the system's outputs among users (for example, government agencies, policymakers, and development partners).
- *A system inherits multiple-donor support.* The existence of multiple donors and a sufficient level of resources strengthens the perception of the SAKSS as a general public good in support of local interests and capacity needs.
- *A system should build credibility and trust among competing partners.* A SAKSS network is not the only player in providing strategic analysis and information for informing ARD strategies. It is therefore important to maintain a degree of transparency in reaching out to other experts who may have a comparative advantage in a particular area. A SAKSS should use this expertise and have sufficient incentive structures in place for collaboration. This could also take the form of organizing seminars to encourage broader participation by experts outside the network. SAKSS should refrain from giving the impression that its network of partners has sufficient expertise in all areas.

Of course there is no single blueprint for a country SAKSS. The experiences of existing programs highlight the unique conditions in each country with respect to stakeholder needs, human and institutional capacity, the current stock of knowledge, funding levels, data availability and quality, and existing relationships among government entities, donors, and the research community. We have emphasized the many factors that can influence the effectiveness of a SAKSS for promoting evidence-based dialogue and decisionmaking, including the political context, external influences, and relationships among individual champions and their organizational links. These ultimately shape the SAKSS of each country with respect to its governance and institutional structure, relationships with local partners, and analytical agenda, for instance. Despite these factors, however, we laid out some basic principles, definitions, and objectives underlying the SAKSS concept and the process of establishing a country SAKSS. We also offered a step-by-step guideline for setting one up, drawing on the experiences of existing efforts and lessons from the literature.

Finally, the operational aspects of a SAKSS offer a real world opportunity to test the concept and its principles. For example, certain institutional and political economy issues, including individual and organizational interactions, emerge out of the collaboration and networking inherent in a SAKSS. From these issues, a number of important questions arise: What drives the interactions in such networks? What factors constrain their ability to function well (such as incentives, institutional affiliations and tensions, transaction costs, competitiveness, and different underlying development paradigms, values, and approaches)? Does the type of membership mix in the networks affect the credibility of the analysis? Other challenging questions that can arise and are worth exploring include these: How can a SAKSS balance the supply of credible information (which is limited) with its demand (which is almost endless)? Can those who seek the information most avidly also pay for it? If not, what are the trade-offs for accepting external donors' involvement and influence?

We hope this conceptual background and technical guide can serve the purpose of guiding the setup of other country SAKSS networks as the demand for them arises. Already, the technical guide offers a useful guideline for the regional ReSAKSS nodes should they be asked to assist with establishing a country SAKSS network as part of the broader CAADP implementation process.

Part III: Appendixes of Tools and Resources

Appendix A: Tools for Analyzing Strategic Priorities and Impact

The purpose of this appendix is to describe some of the economic tools and approaches that are useful for analyzing strategic policy and investment priorities and for impact assessment. For each approach, we review (1) its main purpose; (2) the questions it can help to answer; (3) the conditions under which it should be used; (4) its requirements in terms of data, skills, and cost; and (5) its limitations. For each approach, we also provide a few examples of its application.

Computable General Equilibrium (CGE) Model

Table A.1 answers some basic questions about this model.

Contributions have been made to this chapter by Sam Benin, research fellow, Development Strategy and Governance Division, International Food Policy Research Institute.

Table A.1. CGE model

Question	Answer
What is its purpose?	To capture the interlinkages among sectors and institutions in the economy and address questions relating to public expenditures and budgeting. A CGE model treats both agricultural and nonagricultural sectors with the same level of detail.
What questions can it help to answer?	What is the role of agriculture in the economy? How does it contribute to economywide growth and national development goals?
	What agricultural growth rates would be needed to reach the Millennium Development Goals?
	What sector investments are needed?
	How would changes in demand-side constraints (such as marketing opportunities, transportation costs, and other factors) influence pro-poor growth?
	What are the interactions among the agricultural and nonagricultural sectors, as well as among the rural and urban areas, in the growth process?
	How does growth in each of the major economic sectors and subsectors influence poverty reduction?
	How do the links between economic growth and poverty reduction vary among different regions?
	Are there opportunities for regional linkages that would generate spill-ins or spillovers in the areas of trade and technology?
Under which conditions should it be used?	When it is necessary to capture both direct and indirect effects of policy changes on poverty and income distribution.
What are its requirements?	*Data/information:* Requires a social accounting matrix constructed from national accounts and survey data.
	Time: 3 months–1 year.
	Skills: Advanced modeling experience.
	Supporting software: Excel, Eviews, Gauss, GAMS.
	Financial cost: Depending on the availability of data and whether a SAM has already been built, US$25,000–US$75,000.
What are its limitations?	Cannot be used to assess monetary policies, such as changes in interest rates, inflation, or money supply. Requires extensive data.
Where has its application been documented?	Diao et al. (2007a).

Conceptual framework and methodology

Various analytical tools can be used to assess the impact of policies on *either* growth or poverty. A key challenge when formulating policies, however, is to understand how those policies *simultaneously* affect growth and poverty. Developing such an understanding is difficult because it entails accounting for the actions and interactions of many agents within the economy, including producers, households, enterprises, and governments. Each of these agents has distinct interests and operates in a unique environment. For example, rural households behave differently and have different opportunities than do urban households, and agricultural producers use different technologies and face different constraints than do manufacturing producers. Therefore, a method is needed that takes into account the unique characteristics of and agents within the country under consideration.

Understanding how policies affect growth and poverty is further complicated by the many linkages among economic agents. These linkages determine how the immediate effects of policies on some agents trickle down through the economy and affect others. For example, expanding growth in urban-based agroprocessing may stimulate demand for agricultural inputs, which in turn may generate employment for rural households. Hence, expanding agroprocessing directly benefits urban households while indirectly benefiting rural households. Accounting for linkages is also important in public investment. Although many studies consider how increasing government spending on education or health will affect poverty, few consider the implications of financing these policies. For example, if the government must raise taxes to fund a policy, by how much should they be raised, who will be affected, and how will other areas of the economy be affected? As illustrated in these examples, using analytical methods that consider only the direct effects of policies may miss many of the indirect effects, with potentially disastrous consequences.

In summary, what is needed is an analytical tool or economic model that accounts for the various economic actors within a country. Such a model has to capture the entire economy (that is, it must be economywide). Furthermore, the model should not ignore the many linkages among the various actors. In other words, the model must consider how the macroeconomics of production and growth affect the microeconomics of employment and poverty (that is, it must be a "macro-micro" model). The CGE model is the prime example of an economywide macro-micro model.

The CGE model has two parts. The first part is a purpose-built database called a social accounting matrix (SAM). The data are drawn from a wide range of sources, including household surveys, national accounts, input–output tables, trade data, and so on. The SAM must be constructed before developing the second part of the CGE model, which is a set of mathematical equations that describes the behavior of and linkages among the various economic agents in the country's economy. These equations are then used with the data (in a process called calibration) to ensure that the structure and behavior of the model capture the unique characteristics of the country's economy (see Figure A.1 for an example of the various agents in the CGE model and some of the linkages among them).

In building a CGE model, the challenge is to link economic growth (the left side of Figure A.1) with household incomes and poverty (the right side of Figure A.1). Unlike typical macroeconomic models, the CGE model can be more detailed in its treatment of how economic growth is generated because it includes many disaggregated sectors (drawing on input–output tables and national accounts). This disaggregation is important because policy analysts want to understand the sector-specific technologies used by individual producers (for example, their factor inputs and productivity) as well as the linkages among sectors (for example, cross-sector demands for intermediate goods).

Economic growth is achieved by exploiting factors such as labor, land, and other assets to produce goods and services. However, the factors used and goods produced vary from sector to sector. The CGE model takes these details into account by including many disaggregated factors and commodities (drawing on household surveys). For example, the CGE model has the capacity to differentiate between skilled and unskilled labor, male and female labor, and agricultural and nonagricultural capital. Such distinctions are important because, although employment may generate factor returns, it is important to know the distribution of factor or asset endowments across households in the country. For example, some households do not have access to land and therefore are less likely to benefit from policies that improve returns to land. Other households may not have skilled labor and therefore are unlikely to participate in skill-intensive sectors. Therefore, the income and expenditure patterns of households will depend on their asset endowments, as well as their demographic, geographic, and economic conditions. There are also many linkages among households, especially rural and urban households, through migration and remittances. The CGE model will capture the differences and linkages among households, drawing on household surveys, to allow an assessment of the effects of policies on growth, poverty, and inequality within a single consistent analytical framework.

The CGE model captures the effects of trade by allowing producers to supply both domestic and foreign markets and by allowing import competition once domestic demand is satisfied. Apart from changes in trade policies, the CGE is also

Figure A.1. Overview of the agents and linkages underlying a computable general equilibrium model

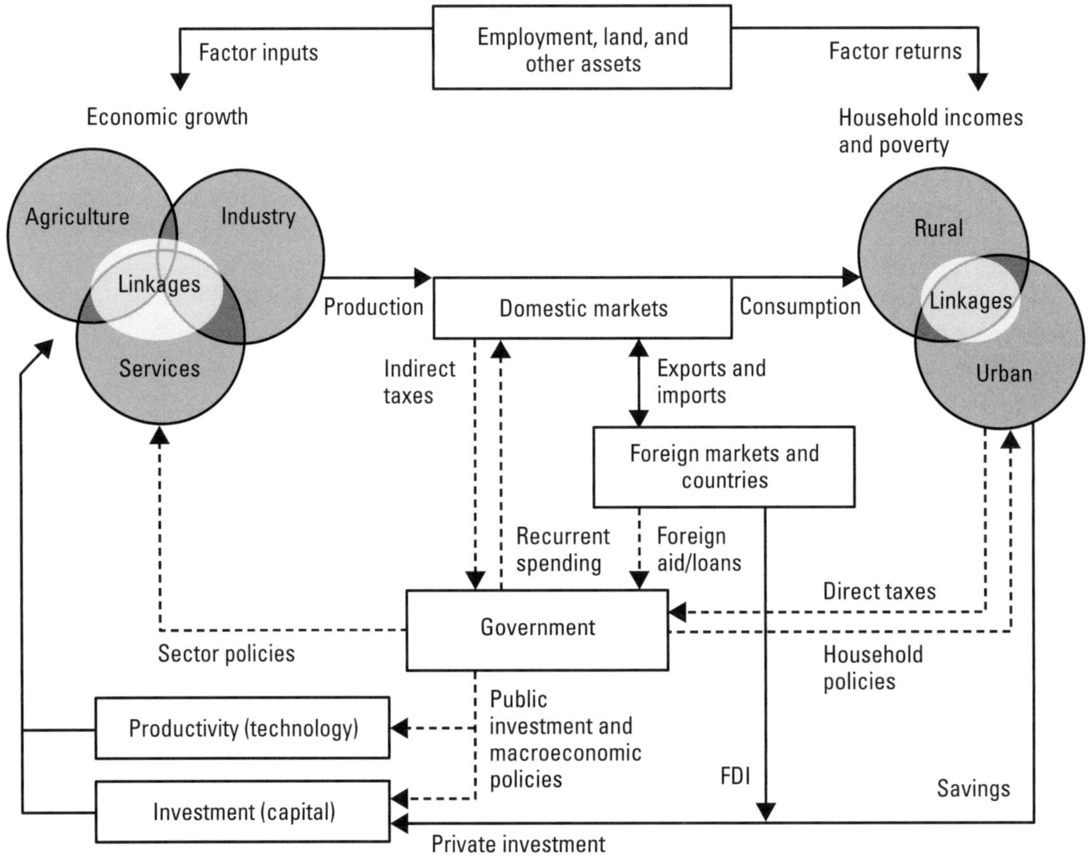

Source: Benin et al. (2008b, Figure 3).

Note: FDI means foreign direct investment.

able to assess the impact of changes in a country's terms of trade on growth and poverty, such as an increase in world oil prices.

The CGE model is a useful tool for assessing a wide range of policies because it captures many of the primary functions and constraints of the government (see dotted lines in Figure A.1). For example, one of the main advantages of using a CGE model is that it can assess how public investment and macroeconomic policies affect growth and poverty, as well as how these policies can be financed. This is particularly important in determining whether there are trade-offs or synergies among different policies. The model will capture the effects of investment on productivity and the accumulation of capital, as well as how such effects, in turn, influence growth in different sectors. In addition, the model will enable an assessment of the effects of direct sectoral policies, such as targeted subsidies, as well as household policies, such as state pensions and other transfers. In terms of financing these policies, the model captures a wide range of tax instruments (that is, personal, corporate, sales, and import taxes) as well as alternative sources of funds, such as foreign aid and borrowing. Together, these features make the CGE model a powerful tool for informing policymakers and ensuring consistency between macroeconomic and poverty reduction strategies.

It should be noted that although the CGE model is one of the more comprehensive analytical tools available, it cannot answer all questions on its own. For instance, this type of CGE model cannot be used to assess monetary policies, such as changes in interest rates, inflation, or the money supply. Such issues usually require a macrofinancial model. Potential complementarities with other policy tools could be explored in the future, but the focus on developing informed and consistent data for growth and poverty reduction policy makes the CGE model an appropriate first step.

ECONOMYWIDE MULTIMARKET (EMM) MODEL

Table A.2 answers some basic questions about this model.

Table A.2. EMM model

Question	Answer
What is its purpose?	To identify the role that the rural and agricultural economy can play in achieving growth and poverty reduction through its linkage with the nonagricultural sector and with the overall economy.
What questions can it help to answer?	Is there a trade-off between growth and poverty reduction?
	Is there a trade-off between growth and inequality?
	Are the sectoral growth targets from national development strategies consistent with overall economic growth and poverty reduction goals?
	In which sectors is growth the most pro-poor?
	What are the sources of growth within agriculture?
	Which subsectors' growth is more pro-poor within agriculture?
Under which conditions should it be used?	When data are too limited to use the more complex CGE model. The EMM model does not disaggregate nonagricultural sectors.
What are its requirements?	*Data/information:* Household income or consumption data.
	Time: 1–3 months.
	Skills: Previous modeling experience.
	Supporting software: Stata, SAS, GAMS.
	Financial cost: US$15,000–US$45,000.
What are its limitations?	Cannot capture interlinkages between agriculture and nonagriculture through factor mobility (especially employment and migration) or backward and forward input–output linkages in production.
Where has its application been documented?	Omamo et al. (2006); Al-Hassan and X. Diao (2007); Diao and Nin-Pratt (2007); Diao and Sarpong (2007); Diao et al. (2007b); Breisinger et al. (2008).

Conceptual framework and methodology

The EMM model captures the detailed structure of the agricultural sector. By contrast, the disaggregation of the CGE models is more evenly balanced across agricultural and nonagricultural sectors. Although the CGE models are better at capturing cross-sector growth linkages during the production process, they provide less detailed information on agriculture, such as the production technologies used in the many agricultural subsectors. These differences may not prove too severe, however, because consumption linkages outweigh production linkages in most developing countries during their early stages of development (Vogel 1994). The ability to capture detailed consumption linkages depends largely on the disaggregation of households' income and expenditure patterns. In this regard, both EMM and CGE models should have highly disaggregated representative households and be linked directly to household expenditure surveys to ensure that the most detailed household information is retained. The models can be disaggregated further across regions within each country to capture the geographic heterogeneity of sectors and households.

The EMM model is based on neoclassical microeconomic theory. In the model there are aggregate producers representing subnational production in each sector and for rural and urban areas. The supply functions are derived under producer profit maximization based on the producer prices of all commodities. In the agricultural sector, these supply functions have two subcomponents: (1) yield functions and (2) land allocation functions responsive to changing profitability across different crops given the total available land. Where data are available, the supply functions are further disaggregated across technologies. For example, the yield functions for farmers employing modern inputs, such as fertilizer or improved seeds, have higher productivity coefficients than those for farmers not using modern inputs. The area functions for each crop are also disaggregated according to the use of modern inputs. For example, the area functions for maize production include farmers using fertilizer only, those using fertilizer and improved seeds, or those not using any modern inputs. Further, because irrigation is treated as one of these modern inputs, there are different supply functions for irrigated and rainfed crop production.

Representative consumers are aggregated from household survey data to represent an average household's consumption patterns in rural and urban areas of each subnational region. The demand functions are derived from utility maximization based on prices and income. Income is generated from both agricultural and nonagricultural activities and is an endogenous variable that links supply with demand as in a typical general equilibrium model.

As the name of the model suggests, a multiple market structure is specified. There is perfect substitution between domestically and internationally produced commodities. However, transportation and other market costs distinguish trade in the domestic market from imports and exports. For example, although imported maize is assumed to be perfectly substitutable for domestically produced maize in consumers' demand functions, maize may still not be profitable to import if its domestic price is lower than the import parity price less any transactions costs. Maize imports can occur only when domestic demand for maize grows faster than domestic supply and the local market price rises significantly. A similar situation applies to exported commodities. Even though certain horticultural products are exportable, if domestic production is not competitive in international markets, owing to either low productivity or high transaction costs, exports will not be profitable. Only when domestic producer prices plus market costs are lower than the export parity price of the same product does it become profitable to export.

The model does not capture bilateral trade flows across subnational regions, although it does identify subnational regions as being food surplus or deficit areas by comparing regional-level demand and supply for total food commodities. Although producers and consumers in different regions operate in the same national markets for specific commodities, prices can vary across regions owing to differences in transportation and market costs. For example, in the Ethiopian EMM model described in Diao et al. (2007a), domestic marketing margins are defined at the regional level according to the distance to Addis Ababa, which represents the central market for the country. For a food surplus region, foodcrop prices faced by local producers are equal to the prices in the central market subtracting market margins, whereas for a food deficit region local prices are higher than those in the central market owing to marketing margins.

To analyze the growth–poverty effect, the nationally defined poverty line is adopted in the model rather than using the World Bank's dollar-a-day measure. National poverty lines are typically measured by total household expenditure because household income is often significantly underreported in developing countries. However, changes in the representative households' expenditures in the model are the results of changes in their incomes (that is, both expenditures and incomes are endogenous variables in the model).

A microsimulation model is used to capture detailed household consumption patterns. More specifically, all sampled households in the household living standard and monetary survey are used to construct the microsimulation model that links with their corresponding representative consumers in the EMM model, who in turn are defined across subnational regions and for both rural and urban areas. A top-down linkage is defined from the EMM model to the microsimulation model. For example, if the results from the Ethiopian EMM model find that a 1 percent increase in GDP causes a 1.3 percent increase in annual spending on teff for the household in the EMM model representing the Ethiopian region "rural West Tigray," there will be 1.3 percent increase in spending on teff by each of the 143 sample households it represents in the microsimulation model. However, the share of teff in each of the 143 households' total expenditure varies. Therefore, a 1.3 percent increase in teff expenditure will affect each of the 143 households differently depending on the share of teff in their consumption basket. As a result, the effect on total household expenditures will be larger for households spending a larger share of their income on teff. These differential effects occur across all commodities in the EMM model. According to these differential effects, the EMM microsimulation models are able to estimate national or regional income distribution changes.

In general, because of the larger share of staple food in poor households' budgets, the same income elasticity for all rural households can result in different aggregate effects on total expenditures across households. Given a fixed poverty line defined by real expenditures, some poor households whose per capita expenditures are initially below the poverty line may move out of poverty in certain years if their expenditures rise above the poverty line. Using the microsimulation model, the national poverty rates are recalculated according to updated total expenditures for each sample household for each year in a simulation.

International Model for Policy Analysis of Agricultural Commodities and Trade (IMPACT)

Table A.3 answers some basic questions about this model.

Table A.3. IMPACT

Question	Answer
What is its purpose?	To project the future of global food production and food market performance and the impact of long-term climate change as measured by water availability.
What questions can it help to answer?	How are the following projected to change in the future: crop area, yield, production, consumption, prices, and trade; livestock numbers, yield, production, consumption, prices, and trade?
	How will the role of agricultural commodities change and impact food security and rural livelihoods?
	What will be the water constraints at the regional, national, and subnational levels?
Under which conditions should it be used?	The minimum scale should be national, with more robust analysis occurring at larger regional and watershed scales.
What are its requirements?	*Data/information:* Production, trade, and consumption data; macroeconomic data; estimates of supply and demand systems (elasticities); estimated growth trajectories.
	Time: Depends on level of analysis.
	Skills: Microeconomics; hydrological engineering; resource economics; database management (collection, cleaning, maintenance, updating); computer programming; commodity and market analysis.
	Supporting software: Excel, GAMS, statistical software.
	Financial cost: IMPACT-D is freely available and requires only a PC; the full IMPACT-WATER (described in "Conceptual framework and methodology") is embedded in IFPRI projects and therefore difficult to cost.
What are its limitations?	It is a partial equilibrium framework focused on the agricultural sector and therefore must use exogenous assumptions for more macro-scale trends in labor, capital, and sector growth.
Where has its application been documented?	Rosegrant et al. (2001).

Conceptual framework and methodology

Although improvements in agricultural production have occurred and malnutrition has declined at the global level since the 1960s, serious questions remain about the ability of world agriculture to continue to realize significant increases in developing-country food availability into the 21st century. With suitable arable area throughout much of the world already under crop production and the advancements of the Green Revolution essentially complete, agricultural research and development will be an increasingly crucial variable affecting future food production increases, and contributing to improved food security. IMPACT can address these questions and examine alternative futures for global food supply, demand, trade, prices, and food security.

IMPACT covers 32 commodities (which account for virtually all of world food production and consumption), including all cereals, soybeans, roots and tubers, meats, milk, eggs, oils, meals, vegetables, fruits, sugar and sweeteners, and fish in a partial equilibrium framework. It is specified as a set of country-level supply and demand equations in which each country model is linked to the rest of the world through trade. In order to explore food security effects, IMPACT projects the percentage and number of malnourished preschool children (0 to 5 years old) in developing countries as a function of average per capita calorie availability, the share of females with secondary schooling, the ratio of female to male life expectancy at birth, and the percentage of the population with access to safe water.

A wide range of factors with potentially significant effects on future developments in the world food situation can be modeled based on IMPACT. They include population and income growth; the rate of growth in crop and livestock yield and production; feed ratios for livestock, agricultural research, irrigation, and other investments; commodity price policies; and elasticities of supply and demand. For any of these underlying factors, IMPACT generates projections for crop area, yield, production, demand for food, feed and other uses, prices, and trade and, for livestock, numbers, yield, production, demand, prices, and trade.

The model is revised periodically to update the base year, include additional commodities, and improve its regional disaggregation capabilities. During 1998–2000, the IMPACT-WATER model was developed, which combined the IMPACT model with the Water Simulation Model (WSM) in order to estimate the interactions between water supply and demand and food supply, demand, and trade. New advancements for the suite of IMPACT models (including IMPACT and IMPACT-WATER) include a greater level of spatial disaggregation, the inclusion of additional crops, and an update of the base year to the year 2000. Although the primary IMPACT model divided the world into 36 countries and regions, the IMPACT-WATER model uses a finer disaggregation of 69 basins in recognition of the fact that significant climate and hydrologic variations within regions make the use of large spatial units inappropriate for water resource assessment and modeling.[34]

[34] For further information or to download a version of the model, go to <http://www.ifpri.org/dataset/dream-dynamic-research-evaluation-management> (accessed September 30, 2010).

SPATIAL ANALYSIS OF DEVELOPMENT OPTIONS

Table A.4 answers some basic questions about this process.

Table A.4. Spatial analysis of development options

Question	Answer
What is its purpose?	To identify and map the magnitude and distribution of opportunities and challenges within and across countries in order to better target appropriate development alternatives.
What questions can it help to answer?	What are the extent and distribution of poverty and food insecurity?
	What are the spatial patterns of livelihood options and constraints?
	How does agricultural potential correspond to population density and market access?
Under which conditions should it be used?	When the geographic variability of the area would influence the choice of policies. Consultations with key stakeholders are required to verify that the domains correspond to local realities and to consider other information not available through data alone.
What are its requirements?	*Data/information:* Agroclimatic data, land use data, production data, market and trade data, household census surveys.
	Time: 1–6 months.
	Skills: Familiarity with geographic information systems (GIS) analysis.
	Supporting software: ArcView.
	Financial cost: Depending on availability of data and expertise, US$20,000–US$100,000.
What are its limitations?	Often data are not georeferenced but are summarized at the district or provincial level.
Where has its application been documented?	Chamberlin, Pender, and Yu (2006); Omamo et al. (2006).

Conceptual framework and methodology

Useful tools in the spatial analysis of development domains are GIS and remote sensing. These tools typically involve overlaying several spatial maps that examine environmental and land use systems in order to highlight any correlations that may exist among them and across space (Dalal-Clayton and Bass 2002). Although quite effective in influencing policies, these tools can easily be misleading if they are not integrated with other more sophisticated behavioral or normative models. Simple correlations do not go far enough to address the socioeconomic interrelationships that are so often relevant to policymaking. When combined with socioeconomic data and analysis, GIS can provide a powerful way to communicate the results of more complex interrelationships (Yeh 1999).

One way to spatially disaggregate the range of rural livelihood options is to define "development domains" that represent a unique combination of key factors, such as land use, farming practices, and income sources, that influence the types of options available at the community level. In Uganda, for instance, Pender et al. (2001) define development domains on the basis of population density, access to markets, and agroclimatic conditions. There are strong associations among these three factors and the basic types of livelihood strategies pursued by most communities, including expanded cereal crop production, intensive livestock/dairy production, agro-processing, and non-agriculture-based rural enterprises.

In different local circumstances, other factors, such as socioeconomic conditions, resource endowments, and vulnerability to production shocks, will also have important implications for characterizing the livelihood options available. In some instances, focusing more attention on commercialization issues may be more critical, requiring domain specifications that are distinguished by existing regional and global end markets for exports and domestic markets, such as large urban centers, for food staples. By adding aspects of market accessibility to the analysis, commodity-specific domains can be mapped out that characterize how end markets are linked to certain attributes along a supply chain.

Ultimately, data limitations will most likely define the type and degree of disaggregation possible. A difficulty inherent in disaggregating socioeconomic data by domain is that many of the data are not georeferenced but are summarized at the district or provincial level. In reviewing some of these types of problems, Wood and Chamberlin (2003) offer possible solutions that involve combining satellite image data with mathematical techniques such as maximum entropy. However, from a purely practical perspective, it is important to consider whether conducting the analysis at this higher resolution will provide information that is more valuable than that produced at a more aggregate level. For example, it may be preferable to analyze sectorwide or thematic issues (for example, the physical infrastructure and policy environment) at the level of the economy if the costs of doing so are exorbitantly high at the domain level.

Because the "development domain" characterization can ultimately shape the type of policy intervention and public investment alternatives available to policymakers, consultation with key stakeholders is required not only to validate the domains to local realities but also to consider other information not available through data alone. Moreover, doing so ensures that any further analysis performed by domain will have local relevance and legitimacy among government decisionmakers and those in the broader development community (for example, practitioners, NGOs, researchers, private-sector groups, and donors).

Econometric impact analysis of public investments

Table A.5 answers some basic questions about this process.

Table A.5. Econometric impact analysis of public investments	
Question	**Answer**
What is its purpose?	To understand the marginal effect that a unit of investment has on a specific outcome variable, such as growth or poverty.
What questions can it help to answer?	What have been the trends of government expenditures by sector, and what have been the reasons for their changes?
	How has public investment been financed, and how has the burden of financing investment policy been distributed in society?
	What have been the returns to various types of government expenditures in terms of their growth, poverty reduction, and other human development effects?
	How can public investment be used to strengthen weak links between poverty reduction channels so as to increase efficiency in targeting public resources at poverty reduction?
Under which conditions should it be used?	When guidance is needed on sectoral spending and when time-series expenditure data are available and reliable.
What are its requirements?	*Data/information:* Time-series expenditure data, household surveys.
	Time: 1–6 months.
	Skills: Experience with econometrics.
	Supporting software: Excel, Stata.
	Financial cost: Depending on availability of data, US$10,000–US$75,000.
What are its limitations?	Reliable expenditure data can be difficult to obtain. It may be difficult for the model to capture the indirect effects of public investment or account for unpredictable exogenous shocks.
Where has its application been documented?	Fan, Hazell, and Thorat (1999); Fan, Zhang, and Zhang (2002); Fan, Zhang, and Rao (2004).

Conceptual framework and methodology

Public investment affects rural poverty through many channels (Figure A.2). A key underlying assumption is that public and private investments are complements (Anderson et al. 2006), so an increase in the public capital stock raises the productivity of all factors in agricultural production, which in turn leads to higher farm wages and incomes and to poverty reduction. For example, public investment in agricultural research, rural education and health, and infrastructure increases farmers' income directly by increasing agricultural productivity, which in turn reduces rural poverty. Indirect effects come from higher agricultural wages and improved nonfarm employment opportunities induced by growth in agricultural productivity. Agricultural output from rural investment often yields lower food prices, again helping the poor indirectly because they are often net buyers of foodgrains. Redistribution of land caused by greater agricultural growth also has important effects on rural poverty. In addition to their productivity effect, public investments in rural education, health, and infrastructure directly promote rural wages, nonfarm employment, and migration, thereby reducing rural poverty. For example, improved infrastructure access will help farmers set up small rural nonfarm businesses such as food processing and marketing enterprises, electronic repair shops, transportation and trade businesses, and restaurant services.

Investments in the rural sector not only contribute to growth, employment, and wages in rural areas but also help the development of the national economy by providing labor, human and physical capital, less expensive food, and markets for urban industrial and service development. Growth in the national economy reduces poverty in both rural and urban sectors. Understanding these different effects provides useful policy insights to improve the effectiveness of government poverty reduction strategies. In particular, it provides information on how public investment can be used to strengthen weak links among poverty reduction channels to increase efficiency

Figure A.2. Government spending and poverty

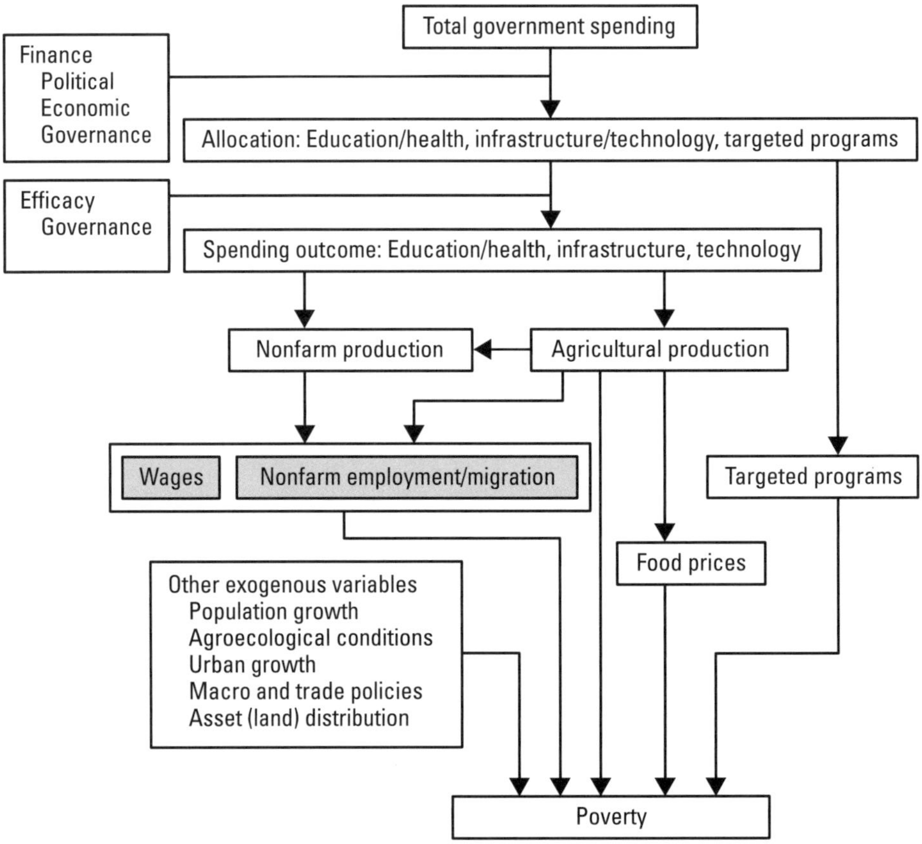

Source: Fan (2008).

in targeting public resources at poverty reduction. More efficient targeting has become increasingly important in an era of macroeconomic reforms in which governments are under pressure to reduce budgets.

There are several challenges in estimating the overall effects of different government spending on growth and poverty reduction.

Endogeneity and reverse causality

Government investment may itself be an endogenous variable. Binswanger et al. (1989) argue that government may allocate its investment based on agroclimatic conditions; that is, high-potential areas may receive more resources from government. If this is true, a simple ordinary least squares regression technique may result in biased estimates. In this case, the return to public investment in terms of growth may be overstated. On the other hand, if the government targeted its resources to poor areas for poverty reduction purposes, the poverty reduction effect may be understated if the endogeneity problem is not properly dealt with. These biases may vary by region and by type of investment.

Similarly, the existence of reverse causality between government investment and development outcome may also result in biased estimates if it is not taken into consideration. Reverse causality occurs because income growth may increase the demand for infrastructure or other public capitals.[35] However, more infrastructure or other public capitals may also induce increases in income.

Besley and Case (1994) argued that endogeneity can also be a result of political and economic factors, which vary over time as well as space. In this case, the fixed-effects approach used by many economists does not resolve the endogeneity problem because it fails to control for the omitted time-varying differences across space, which help to determine policy and outcome (Van de Walle 1998).

One of the most common approaches to avoid the potential biases in the estimates due to endogeneity and reverse causality is the instrumental variable approach. Broadly

[35] For more information on reverse causality, see Kessides (1993); World Bank (1994); Canning and Bennathan (2000); and Zhang and Fan (2004).

speaking, an instrumental variable is a variable that is uncorrelated with the error term but correlated with the explanatory variables in the model. But in reality it is hard to find such an instrument (or instruments). Davidson and MacKinnon (1993) demonstrated that the validity of the choice of instruments may be tested in this context via an auxiliary regression.

When panel data are available, the two-way fixed-effects model can eliminate most of the biases due to time- or regional-invariant fixed effects. For example, if government always targets its resources to a particular region (for example, either a high-potential or a poor region), a regional fixed-effects model should be able to eliminate the endogeneity bias.

However, an understanding of government budget allocation and political economy may be fundamental to public spending effects (Pitt, Rosenzweig, and Gibbons 1993; Besley and Case 1994; Van de Walle 1998), which is addressed in the next section.

Multiple effects and simultaneous equations

Most of the previous studies estimating the effect of public investment on growth and poverty have used a single-equation approach. There are at least two disadvantages to this method. First, many poverty and inequality determinants, such as income, production or productivity growth, prices, wages, and nonfarm employment, are generated from the same economic process as inequality and poverty. In other words, these variables are also endogenous variables. Ignoring this characteristic leads to biased estimates of the poverty and inequality effects.

Second, certain economic variables affect poverty and inequality through multiple channels. For example, improved rural infrastructure reduces rural poverty not only through improved growth in agricultural production but also through improved wages and opportunities for nonfarm employment. It is very difficult to capture these different effects using a single-equation approach.

As discussed, to avoid the endogeneity problem, the reduced form is often used, and various instruments have been used to measure the explanatory variables, which are potentially endogenous. But to solve the second problem, a structural equation system is needed.

For the past several years, IFPRI has been developing a simultaneous equations model to estimate the various effects of government expenditure on production, inequality, and poverty through different channels (Fann, Hazell, and Thorat 1999; Fan, Zhang, and Zhang 2002; Fan, Fang, and Zhang 2003). The equations in the model are specified based on economic behavior and have been used by various scholars in the single-equation approach. The empirical model developed mimics the rural economy where government spending is the driving force behind rural poverty reduction, controlling for other factors. The first equation measures the effects of growth in agricultural productivity, improvement in nonfarm employment and wages, and changes in terms of trade (changes in agricultural prices relative to nonagricultural prices) on reduction in rural poverty. The second equation is an agricultural productivity function in which growth in agricultural productivity is a function of government investment in agricultural R&D and public capital stocks of irrigation, power, education, and rural roads. The third and fourth equations capture the effects of growth in agricultural productivity and increased public capital stock such as roads, education, and power on rural wages and nonfarm employment. The next set of equations models how different government expenditures form capital stocks in education, irrigation, roads, and other types of public capitals. Finally, agricultural prices are modeled as functions of agricultural productivity growth because increased supply due to growth in agriculture will depress agricultural prices, indirectly affecting rural poverty.

In addition to its ability to track the relevant linkages among public investments and rural poverty, a systems approach enables other endogenous variables to be properly specified. Once the model is estimated, the total effects of public investment variables on growth and poverty reduction can also be calculated by totally differentiating the equations system with respect to each public investment variable.

The major disadvantage of this approach is its requirement of correct specifications of all equations. Even when one equation is misspecified, the bias will affect the estimated coefficients in all equations in the system. To ensure the reliability of the estimated results, various specification tests should be performed (Greene 1993).

Time lag of investment

Government investments in R&D, roads, education, power, health, and irrigation can have long lead times in affecting agricultural production, as well as long-term effects once they kick in. One of the thornier problems to resolve when including government investment variables in a production or productivity function concerns the choice of an appropriate lag structure.

Most past studies used stock variables, which are usually weighted averages of current and past government expenditures on certain investments such as R&D. But what weights and how many years' lag should be used in the aggregation are currently issues of some contention in the literature.[36] When

[36] Alston, Norton, and Pardey (1995) argue that research lag may be much longer than previously thought, or even infinite. But in many developing countries the National Agricultural Research Systems are much younger than those in developed countries (often 30 to 50 years old), and their research is of more applied types. Therefore, it is certain that research lags in developed countries are much shorter than those in developing countries.

there is lack of long time-series data on government investment (by types and regions), the stock approach can be used as a crude proxy. Some of the sensitivity analysis from China and India shows that the ranking of different public investments in terms of their growth and poverty reduction effects changes very little, but the magnitudes of the effects may change.

However, when the long time-series data on investment are available, the following procedures can be used to determine the lag structure or the dynamic relationship between government investment and the final development outcome. A first step is to use statistical tools to test and determine the appropriate length of lag for each investment expenditure. For example, include annual agricultural research expenditures for a certain number of past years in the agricultural production function. How many years should be included depends on the statistical test value. Various procedures have been suggested for determining the appropriate lag length. The adjusted R^2 and Akaike's information criteria (AIC) are used by many economists (Greene 1993). The optimal length is determined when the adjusted R^2 reaches its maximum or AIC reaches the minimum.

However, the coefficients of the annual past government expenditures cannot be used directly in calculating the effects on growth in agricultural production because these variables are often highly correlated, making the estimated coefficients statistically insignificant. To avoid this problem, the most popular approach is to use what are called polynomial distributed lags. For more detailed information on this subject, refer to Davidson and MacKinnon (1993).

Spillover effects

Many types of public investment may have spillover effects, for example, within regions in a country and between different countries. One typical example is agricultural research.[37] Although it is empirically difficult to estimate these effects, the potential gain from consolidating agricultural research systems is large (Byerlee and Traxler 1996).

There are several types of spillover effects. The most common is technology spillover. For example, a new technology developed by a public research institute may be adopted by other regions within the same country or even by other countries. The effects of public investment may also spill over to other regions or even sectors through prices. Increased road investment will increase the food supply in one region, leading to a reduction in food prices not only in its own region but also in other regions or even urban sectors (Fan, Hazell, and Thorat 1999; Fan, Fang, and Zhang 2003). Investment may also affect other regions through the labor market. In a study of the impact of the Green Revolution in a sample of Asian villages, David and Otsuka (1994) found that seasonal migration played an important role in spreading the benefits between technology-adopting and nonadopting regions. Ignoring these spillover effects can lead to either understating or overstating economic benefits and poverty reduction effect.

The impact of spending at higher administrative levels

An issue that has not been dealt with is evaluating spending at higher administrative levels (for example, federal or central) than the unit of analysis (that is, district or region). This is especially the case with defense and R&D spending, among others. A macroeconometric model, which requires a much longer time-series data for estimation, may be needed. Otherwise one has to first make some assumptions about how such federal or central spending is distributed across districts or regions. If thought of in terms of spillover effects, such as information or technology diffusion in the case of R&D, a variable that is constructed as a function of the distance from the center to the region or district can be included in the econometric regression. Nevertheless, given that spending at the federal or central level will unlikely be distributed proportionally to district- or region-specific spending, there is likely to be an obvious bias in ignoring spending at the central level.

Controlling for other factors

Many other factors may affect the development outcome in addition to public investment. These variables may include changes in international trade and prices, domestic macroeconomic conditions, urban development, and regional agroecological conditions. For example, institutional changes and policy reforms have made large contributions to rapid growth in agricultural and nonagricultural production and to poverty reduction in China's rural areas since 1979. In India, market and trade liberalization introduced in the early 1990s also had profound effects on economic growth as well as on poverty reduction. If these variables are not controlled for, the estimated effects of public investment on poverty reduction will be biased, and in many cases returns to public investment will be overestimated. A common practice is to use year and regional dummies to control for year- and region-specific fixed effects.

Prioritizing future government spending

Simulation models can be used to guide improved patterns of investment (differentiated by type of investment, type of region, and packaging and sequencing of investments) to achieve specific agricultural growth, rural poverty, and equity goals. In general, this will provide guidance for a long-term investment strategy in developing countries. The ex post evidence on public investment effects will be used as a basis for modeling

[37] Investments in roads and education have the similar features, although less salient.

possible future returns from different public investments. The parameters estimated can also be used for simulation at the economywide level. Two types of simulations can be conducted. First, the effects of public investment in rural areas can have economywide effects. For example, investment in agricultural R&D affects urban residents through lowered food prices and wages, in addition to affecting rural residents. Second, certain economywide policies affect the allocation of public spending to rural areas. For example, how government revenue is financed and allocated would affect government spending on agriculture and rural areas.

To properly capture spillover effects (discussed earlier) as well as those associated with changes in economywide variables such as employment, wages, and prices, general equilibrium models such as economywide EMM and CGE models become necessary.

DREAM (Dynamic Research EvaluAtion for Management)

Table A.6 answers some basic questions about this process.

Table A.6. DREAM

Question	Answer
What is its purpose?	To evaluate the economic effects of agricultural research and development (R&D). Users can simulate a range of market, technology adoption, research spillover, and trade policy scenarios based on a flexible multimarket partial equilibrium model.
What questions can it help to answer?	What economic returns and reductions in hunger and poverty will likely result from specific investments to overcome production and commercialization constraints?
	What will be the effect on prices and trade if certain technologies are adopted in farmers' fields?
	What will be the economic returns for farmers and consumers?
	What will be the rate of return to investment in agricultural R&D?
	What will be the effect and the magnitude of technological spillover into other regions/ agroecological zones?
Under which conditions should it be used?	DREAM could be used to evaluate the impact of commodity-oriented research and productivity improvement. When data are even too limited to use a multimarket or agricultural sector model, DREAM could be an alternative.
What are its requirements?	*Data/information:* The technology's impact on yield and cost, adoption rate, base-year market data.
	Time: 1–4 weeks.
	Skills: Undergraduate economic major or 3-day training.
	Supporting software: DREAM package.
	Financial cost: Freely downloadable from the IFPRI website.
What are its limitations?	It is a single-commodity model without explicit representation of cross-commodity substitution effects in production and consumption.
Where has its application been documented?	Omamo et al. (2006); Johnson et al. (2008).

Conceptual framework and methodology

DREAM is a menu-driven software package for evaluating the economic effects of agricultural R&D (Wood, You, and Baitx 2001). Users can simulate a range of market, technology adoption, research spillover, and trade policy scenarios based on a flexible multimarket partial equilibrium model.

With DREAM, a range of technology investment, development, and adoption scenarios can be defined and saved in an integrated database. Scenarios are described using market, R&D, and adoption information for any number of separate "regions." Some factors, such as taxes, subsidies, growth rates, and price elasticities, can be specified as constant or as changing over the analysis period. Each region in which production takes place may have its own pattern of technology adoption. After specifying the initial conditions for each region, it is possible to simulate the likely effects of technology development and adoption on prices; on quantities produced, consumed, and traded; and on the flow of economic benefits to producers, consumers, and government.

DREAM handles simple to relatively complex evaluation problems using a standardized interface. A number of market assumptions are possible: a small open economy, a closed economy, vertically integrated farm and postharvest sectors in a single economy, or multiple trading regions. The software also accommodates technology-driven shifts in supply or demand, and users may specify constant or variable shift effects over time in farmers' fields. Significantly, DREAM's multiple-region specification can simulate various technology spillover scenarios in which a technology may be adopted in more than one region. Changes in the pattern of technology spillovers can significantly alter the magnitude and distribution of R&D benefits.

DREAM has been applied to the evaluation of individual projects in a national context as well as to entire commodity sectors at a subcontinental or continental scale. And although it was designed primarily to evaluate options for R&D that are yet to be undertaken (ex ante assessments), DREAM has also been successfully applied to analyzing the effects of past research (ex post assessments).

DREAM was designed to measure returns to commodity-oriented research in an open-economy setting, allowing for price and technology spillover effects between a country in which the research originated and the rest of the world (Alston, Norton, and Pardey 1995). DREAM is a single-commodity model, so there is no explicit representation of cross-commodity substitution effects in production and consumption, but these aspects are implicitly represented by the elasticities of supply and demand. The primary parameterization of the supply and demand equations relies on a set of prices and annual quantities in a defined base period and a set of corresponding price elasticities. The idea is that the linear approximation implied by these elasticities will be good for small equilibrium displacements, such as those implied by single-digit percentage shifts in supply or demand, regardless of the true (nonlinear) functional forms of supply and demand. Small shifts have the added virtue that the cross-commodity and general equilibrium effects are likely to be small, and the total research benefits will not significantly depend on the particular elasticity values used (although the distribution of those total benefits among producers and consumers is sensitive to the elasticity values used).

DREAM parameterization defines the supply and demand curves in the base year to replicate observed market prices and quantities. DREAM also allows for underlying growth of supply and demand to project a stream of shifting supply and demand curves into the future that generates a stream of changing equilibrium prices and quantities in the "without-research" scenario. These without-research outcomes can be compared to "with-research" outcomes, in which a stream of supply curve displacements also incorporates research-induced supply shifts. The research-induced supply shifts are defined by combining an estimate of the maximum percentage research-induced supply shift that could be achieved if the innovation were adopted with an assumed profile of the likely levels of adoption over time.

Finally, measures of producer and consumer surplus are computed and compared between the with-research and without-research scenarios, and these are discounted back to the base year to compute the present values of benefits. In a situation in which there are estimates of the costs of the research that is responsible for the supply shift being modeled, a net present value or internal rate of return can be computed.[38]

---

[38] For further information or to download a version of the model, go to <http://www.ifpri.org/dataset/dream-dynamic-research-evaluation-management> (accessed September 30, 2010).

ESTABLISHING A MONITORING AND EVALUATION (M&E) SYSTEM

Table A.7 answers some basic questions about this process.

Table A.7. Establishing an M&E system

Question	Answer
What is its purpose?	To track progress and performance and assess the effect over time of investments and interventions.
What questions can it help to answer?	Are investments on track to meet development goals? If not, what needs to be altered?
	What are the projected effects (keeping track of those derived from past versus future investments) if investment activities proceed as planned?
	Are these projected effects compatible with the goals and target effects of the existing strategy?
	What policies and types of investments can lead to greater and more sustainable growth as well as greater and better-distributed outcomes and effects?
Under which conditions should it be used?	When a government is committed to effective strategy implementation.
What are its requirements?	*Data/information:* Production, trade, consumption, expenditure, macroeconomic, and household data.
	Time: Can range widely, easily taking up to a year to finalize, depending on the level of complexity, status of strategy development, data availability, and requirements for stakeholder input and validation of the indicators.
	Skills: Basic familiarity with economic theory, project evaluation techniques (for example, cost–benefit analysis), and basic statistics and data measurement.
	Supporting software: Spreadsheets and databases.
	Financial cost: Unknown, but rises with the size and complexity of the M&E system and the degree of stakeholder participation.
What are its limitations?	An M&E system is useful only if carefully designed and actively maintained. An overly complicated system is more costly and unlikely to be effective.
Where has its application been documented?	Benin, Johnson, and Omilola (2010).

Conceptual framework and methodology

Developing an M&E system that will actually be used by stakeholders and decisionmakers is not a straightforward exercise. The challenge, especially for those tasked with developing and implementing such a system, is to collect, analyze, and report information that is timely, credible, and well understood in order to inform decisions about whether to revise priorities or reallocate resources. With this challenge in mind, a simple and intuitive approach is recommended, but one that maintains a sound theoretical framework of causality among inputs, outputs, and outcomes to be measured in the system. The simpler it is, the less costly it will be to develop and maintain, the lower the likelihood of errors in measurement, and the fewer the complexities inherent in the system to interpret the information that comes out of it. A large and disparate set of information (indicators) that ends up not being collected or interpreted has no value to anyone. In reviewing the experience of a wide range of government M&E systems, a recent study by the Independent Evaluation Group of the World Bank (Mackay 2007) summarized this well:

- Overengineering an M&E system is not only wasted effort but can eventually undermine the system.
- Simply believing that an M&E system has inherent value is a typical mistake. The information in the system is valuable only if it is used.

With these basic principles in mind, according to Benin, Johnson, and Omilola (2010), some of the questions one should ask in developing an M&E system are these:

- What should be monitored and evaluated, given the underlying theoretical framework and/or logic that describes the chain of causality among investments (inputs), outputs, and outcomes?

- What is the minimum set of indicators that can be used to tell a credible story of how particular investments and policies are influencing outputs and results? The final selection of the minimum set of indicators will depend on such criteria as data availability, frequency of collection, timeliness, and quality.

- What kind of reporting and frequency are desirable to target local decisionmaking processes?

Several documents have been written on how to design and implement M&E systems, with substantial attention given to the selection of indicators, which are at the heart of any performance evaluation system (for example, IFAD 2002). Although there are different ideas about the criteria to be used in selecting indicators, it is generally agreed that the indicators must be SMART (specific, measurable, achievable, realistic, and timely). For an agricultural strategy, this means that the potential indicators must be the ones that best reflect the range of inputs, outputs, and outcomes associated with the priority investments being implemented. They must also capture critical landmarks along the pathway(s) of impact, that is, between the relevant investments that are put in place and how they can affect agricultural productivity growth and poverty

A useful and practical guide for setting up a results-based M&E systems is the World Bank's ten-step approach (Kusek and Rist 2004). Although it is presented linearly, it does not always have to be considered as such. The steps can be summarized as follows (with more detail available from the original source):

1. Conducting a readiness assessment to build the foundation for an M&E system.
2. Choosing outcomes to monitor and evaluate in a participatory and consultative manner.
3. Setting key performance indicators to monitor progress with respect to inputs, activities, outputs, outcomes, and impacts. This should be an iterative process.
4. Establishing performance baselines—qualitative or quantitative—to establish a starting point with which to compare future changes.
5. Selection of results targets to reach on the way to a longer-term outcome—using the baseline information and expectation of future improvements due to investments.
6. Monitoring for results by collecting quality performance data.
7. Determining the uses, types, and timing of evaluation that make sense.
8. Reporting the findings of M&E.
9. Using the findings in generating and sharing knowledge and learning.
10. Sustaining a results-based M&E system (through demand, clear roles and responsibilities, trustworthy and credible information, accountability, capacity, and appropriate incentives).

Appendix B: Terms of Reference of a Country SAKSS Coordinator/Manager

General Information

Position title:	Country SAKSS Coordinator/Manager
Hiring institution:	Implementing Institution responsible for SAKSS
Reports to:	Country senior representative of implementing institution and head of local host institution (for example, Ministry of Agriculture)

Job Description

[Implementing Institution] seeks to hire a [Country] national to work jointly with the [Implementing Institution] country representative and the [Local Host Institution or the Ministry of Agriculture] in the coordination and implementation of a country SAKSS program. The program aims to enhance local capacity for evidence-based policy and data analysis as well as knowledge and information exchanges in the areas of agriculture and rural development. It is to be established within a local government body (such as the Ministry of Agriculture) to provide a means to better manage existing and new knowledge on the agricultural sector and the rural economy in the country and ultimately strengthen the foundation of evidence for policy formulation and for informing development strategy decisions in general. A SAKSS network will be established to fundamentally consist of local and international data providers, researchers, and analysts working in the country. The network will help provide key data analysis, knowledge and information exchange, and capacity strengthening within the ministry (or other local host institution).

Duties and Responsibilities

- Establish a SAKSS node for agriculture and rural development policy and program analysis, design, monitoring, and evaluation.
- Establish a SAKSS network of data providers, analysts, and researchers.
- Manage the compilation of information on past research and relevant datasets on agriculture and rural development.
- Contribute to data analysis upon demand, including analysis of spatial data, in order to serve the needs of the strategy development process and ongoing dialogue in a timely fashion.
- Contribute to the preparation of policy briefs and reports based on the data analysis, ongoing research of collaborators, and emerging policy issues.
- Perform the day-to-day program management and coordination in close collaboration with the implementing institution, the local host institution, a relevant ministry, and other key local stakeholders.
- Serve as a key SAKSS program liaison and link between the Ministry of Agriculture (or other local host institution) and the scientific community, development partners, private-sector and civil-society groups, and nongovernmental organizations.
- Serve as the champion for SAKSS in the sense of achieving active participation and support from local institutions, government ministries, donors, private-sector investors, farmers' organizations, and research institutions and other clients.
- Promote greater knowledge and data sharing through various media (for example, policy seminars, web-based platforms, news media) and the coordination of dialogue linking policy analysis and decisionmaking.
- Assist with managing program budgets and raising resources and prepare progress reports relevant to diverse stakeholder groups.
- Coordinate research teams to generate new evidence for policymaking.

Qualifications and Experience

- Master's degree in agricultural economics, rural development, statistics, geography, rural sociology, or a closely related field
- Proven skills in building network linkages, promoting information exchange
- Excellent management, interpersonal, networking, and team-building skill
- Experience with the management of databases or library holdings

- Proven skills and ability to work with complex quantitative datasets and experience with geographic information systems
- A high level of computer literacy, particularly with software for database and library management, data analysis, and CD and webpage authoring
- Evidence of strong attention to detail
- Excellent written and spoken English and the main local languages
- A willingness to travel, both in country and abroad

Other desirable qualifications

- Knowledge of and experience in working with the principal public-sector providers of analysis and information in the country
- Experience in research on development issues in the country
- Familiarity with quantitative research techniques if possible
- A holistic and solid knowledge of the country's agriculture and policy environment (government, private sector, NGOs) and its evolution in recent years

Appendix C: A Generic SAKSS Communication Strategy

A SAKSS program provides policy-relevant information and analyses to improve policymaking, track progress, document success, and help countries and regions improve their agricultural strategies. The success of the program depends on how effectively its communication strategy has targeted its audience. Table C.1 summarizes typical audiences, communications goals for each audience, the message a SAKSS program would ideally want to send, and how best to reach each audience with those messages.

Contributions have been made to this chapter by Marcia MacNeil, communications specialist, Development Strategy and Governance Division, International Food Policy Research Institute.

Table C.1. Elements of a generic SAKSS communication strategy

Audience	Goals	Message	Vehicle(s)
Policymakers	Promote evidence-based policymaking, provide timely information and knowledge, and support country strategy implementation without appearing too science-driven	SAKSS data and tools have been created by Africans for Africans in order to meet their development goals.	*Publications:* Policy briefs *Events:* High-level conferences attended by key ministry and donor officials *Media:* Interviews and coverage in local newspapers and on radio or television news
Policy advisers	Build awareness, train them on how to interpret the results of SAKSS tools (for example, investment priorities, spatial analysis), and make them aware of SAKSS data and resources	SAKSS tools and data analysis are resources to be used by all (as a public resource).	*Publications:* Training and explanatory materials *Events:* Workshops led by respected local experts *Web:* Tools, data, and contact information for researchers; blog for two-way communication
Regional economic communities (RECs), the New Partnership for Africa's Development (NEPAD), and the African Union (AU)	Share results among RECs, encourage progress, demonstrate results to NEPAD and AU	SAKSS tools and research are effective for peer and mutual review of progress toward development goals.	*Publications:* Documentation of research to highlight results, credibility *Events:* High-level conferences attended by key ministry and donor officials *Web:* Success stories and summaries of research on website *Media:* Coverage of events, interviews with REC officials
Academic and research institutes	Provide access to research, data, and tools	Our research and tools are freely available for all to use.	*Training:* Use of tools *Publications:* Research reports of results and training manuals *Web:* Tools, data, and contact information for researchers; blog for two-way communication
Private sector/local organizations/donors	Demonstrate effectiveness to raise profile and reputation, credibility of program	SAKSS research and tools are effective and relevant and deserve their support.	*Publications:* Short PR pieces; success stories on website *Events:* Include them in larger, more general conferences *Media:* Any coverage in mainstream newspapers

Appendix D: SAKSS Stakeholder and Partner Institutions and Organizations

African Governments

Includes offices of the president, statistical bureaus, and the various ministries involved in agriculture, rural development, irrigation, water resources, infrastructure, transportation, the environment, natural resources, women's affairs, national planning, and finance.

Government of Ethiopia
Government of Ghana
Government of Malawi
Government of Mozambique
Government of Nigeria
Government of Rwanda
Government of Uganda

African Regional Organizations

African Union (AU)
Association for Strengthening Agricultural Research in Eastern and Central Africa (ASARECA)
Common Market for Eastern and Southern Africa (COMESA)
Economic Community of West African States (ECOWAS)
Forum for Agricultural Research in Africa (FARA)
New Partnership for Africa's Development (NEPAD)
Southern African Development Community (SADC)
West and Central African Council for Agricultural Research and Development (CORAF/WECARD)

African Universities and Research and Policy Analysis, Information, and Knowledge Organizations/Networks

Food and Natural Resources, Policies Analysis Network (FANRPAN)
Policy Analysis and Advocacy Programme (PAAP)
Tegemeo Institute of Agricultural Policy and Development

Ethiopia

Addis Ababa University
Ethiopian Chamber of Commerce
Ethiopian Development Research Institute (EDRI)

Ghana

Institute of Social Science and Economic Research (ISSER)
National Development Planning Commission (NDPC)
University of Ghana, Legon

Uganda

Economic Policy Research Center (EPRC)
Makerere University
National Planning Authority (NPA)

Bilateral Donors

Includes central, country, and regional field offices.
Canada: Canadian International Development Agency (CIDA)
Sweden: Swedish International Development Cooperation Agency (Sida)
United Kingdom: Department for International Development (DFID)
United States: United States Agency for International Development (USAID)

International Organizations and Institutions

Famine Early Warning System (FEWS Net)
Food and Agriculture Organization of the United Nations (FAO)
Foodnet (East Africa)
International Development Research Centre (IDRC)
International Crops Research Institute for the Semi-Arid Tropics (ICRISAT)
International Food Policy Research Institute (IFPRI)
International Institute of Tropical Agriculture (IITA)
International Livestock Research Institute (ILRI)
International Research Centre for Maize and Wheat (CIMMYT)
International Water Management Institute (IWMI)
United Nations Economic Commission for Africa (UNECA)
U.S. universities (Michigan State, Purdue, Cornell)
World Agroforestry Centre (ICRAF)

Multilateral Donors

African Development Bank (ADB)
World Bank

Private-Sector Foundations

Bill and Melinda Gates Foundation

This list is not exhaustive. It is intended to illustrate the kinds of organizations and agencies that a SAKSS program typically partners with to form a network platform.

Appendix E: Glossary of Selected Terms

Data	An unorganized and unprocessed collection of facts from which conclusions may be drawn, for example, "1500 m" (Tadesse 2006, adopted from Bouthillier and Shearer 2002).
Development strategy	A long-term policy and investment plan of action designed to achieve high-end development goals.
Evidence	The results of research and analysis, synthesis and trends, local knowledge and information.
Impact assessment	A particular type of evaluation that aims to determine whether and to what extent a program, policy, or strategy causes changes in the desired direction among a target population or in an environment (Rossi and Freeman 1993).
Information	An aggregation of data (processed data) that makes decisionmaking easier. Information is usually labeled as data made meaningful by being put into a context, for example, "Lowland areas which are below 1500 meters elevation are moisture deficient" (Tadesse 2006, adopted from Bouthillier and Shearer 2002).
Knowledge	1. The combination of information and context in a way that makes it actionable, for example, "Farmers need to harvest rain water to supplement crop water requirements in the lowland areas" (Tadesse 2006, adopted from Bouthillier and Shearer 2002). 2. A defined body of information that might consist of facts, opinions, ideas, theories, principles, and models (or other frameworks) (ICASIT 2003). 3. A person's state of being with respect to some body of information, such as ignorance, awareness, familiarity, understanding, and facility (ICASIT 2003).
Knowledge management	1. The act of making an organization's knowledge stores more accessible and useful (ICASIT 2003). 2. A conscious strategy of getting the right knowledge to the right people at the right time and helping people share and put information into action in ways that strive to improve organizational performance (ICASIT 2003). 3. Systematic, explicit, and deliberate building, renewal, and application of knowledge to maximize an enterprise's knowledge-related effectiveness and returns from its knowledge assets (Beckman 1999, citing Wiig 1997).
Knowledge support system	A network that serves the evidence needs of strategy formulation and implementation. This network includes individuals such as researchers, policymakers, and development practitioners and institutions such as government agencies, universities, research institutes, development organizations, and private and civil-society groups. These individuals and institutions are linked through collaborative research (strategic analysis in the case of SAKSS), capacity strengthening, and dialogue. Through these activities, information, data analysis, and knowledge are compiled, synthesized, and packaged into evidence that is supplied on a timely and reliable basis to be of use during strategy processes.
Knowledge systems	1. In computer sciences, mechanisms for storing, organizing, and retrieving information. 2. In the research community, describes the links among *people* (researchers, policy analysts, development practitioners, decisionmakers, and other stakeholders); *organizations and networks* (policy analysis think tanks, universities, government agencies, nongovernmental bodies, and private-sector organizations); and *evidence* (results of research and analysis, synthesis and trends, local knowledge and information) (Cash et al. 2003).
Network	Formal or informal structures that link actors (individuals or organizations) that share a common interest in a specific issue or a general set of values (Perkin and Court 2005).

Policy	A set of principles or rules to guide decisions and actions. Policies are often legislated and enacted. Several policies, as well as programs, may fit into the larger strategy.
Program	A plan of structured activities or steps to be carried out (or goals to be accomplished).
Strategic analysis	An integrated framework of analysis that helps identify policy and investment options for achieving high-end development goals.
Strategy	A long-term plan of action designed to achieve a particular goal.[39]

[39] Definition adapted from Wikipedia.

References

Adato, M., and R. Meinzen-Dick, eds. 2007. *Agricultural research, livelihoods, and poverty: Studies of economic and social impact in six countries.* Baltimore: Johns Hopkins University Press.

Al-Hassan, R., and X. Diao. 2007. *Regional disparities in Ghana: Policy options and public investment implications.* IFPRI Discussion Paper 693. Washington, D.C.: International Food Policy Research Institute.

Alston, J. M., G. W. Norton, and P. G. Pardey. 1995. *Science under scarcity.* Ithaca, N.Y., U.S.A.: Cornell University Press.

Alston, J. M., T. J. Wyatt, P. G. Pardey, M. C. Marra, and C. Chan-Kang. 2000. *A meta-analysis of rates of return to agricultural R&D: Ex pede Herculem?* IFPRI Research Report 113. Washington, D.C.: International Food Policy Research Institute.

Anderson, E., P. de Renzio, and S. Levy. 2006. *The role of public investment in poverty reduction: Theories, evidence and methods.* ODI Working Paper 263. London: Overseas Development Institute.

AU/NEPAD (African Union/New Partnership for Africa's Development). 2005. *Guidance note for agriculture expenditure tracking system in African countries.* Midrand, South Africa.

Ayuk, E. T., and M. A. Marouani, eds. 2007. *The policy paradox in Africa: Strengthening links between economic research and policymaking.* Trenton, N.J., U.S.A., and Ottawa: Africa World Press and the International Development Research Centre.

Babu, S. C., and P. Pinstrup-Andersen. 1994. Food security and nutrition monitoring: A conceptual framework, issues and challenges. *Food Policy* 19 (6): 218–233.

Babu, S. C., V. Rhoe, A. Temu, and S. Hendriks. 2004. *Strengthening Africa's capacity to design and implement strategies for food and nutrition security.* 2020 Africa Conference Brief 7. Washington, D.C.: International Food Policy Research Institute.

Badiane, O., S. Odjo, and J. Ulimwengu. 2010. *Building capacities for evidence and outcome-based food policy planning and implementation: The example of the Comprehensive Africa Agriculture Development Programme.* IFPRI Discussion Paper 1019. Washington, D.C.: International Food Policy Research Institute.

Bandara, J. S. 1991. Computable general equilibrium models for development policy analysis in LDCs. *Journal of Economic Surveys* 5 (1): 3–69.

Beckman, T. 1999. The current state of knowledge management. In *Knowledge management handbook,* ed. J. Liebowitz. Boca Raton, Fla., U.S.A.: CRC Press.

Bell, C., P. B. R. Hazell, and R. Slade. 1982. *Project evaluation in regional perspective.* Baltimore: Johns Hopkins University Press.

Benin, S., M. Johnson, and B. Omilola. 2010. *Monitoring and evaluation (M&E) system for the Comprehensive Africa Agriculture Development Programme (CAADP).* ReSAKSS Working Paper 6. Washington, D.C.: Regional Strategic Analysis and Knowledge Support System facilitated by the International Food Policy Research Institute.

Benin, S., E. Nkonya, G. Okecho, J. Pender, S. Nahdy, S. Mugarura, E. Kato, and G. Kayobyo. 2007. *Assessing the impact of the National Agricultural Advisory Services (NAADS) in the Uganda rural livelihoods.* IFPRI Discussion Paper 724. Washington, D.C.: International Food Policy Research Institute.

Benin, S., J. Thurlow, X. Diao, C. Lungren, and F. Simtowe. 2008a. *Agricultural growth and investment options for poverty reduction in Malawi.* IFPRI Discussion Paper 794. Washington, D.C.: International Food Policy Research Institute.

Benin, S., A. Nin-Pratt, S. Fan, C. Breisinger, T. Mogues, J. Thurlow, and X. Diao. 2008b. *Growth and poverty reduction impacts of public investments in agriculture and rural areas: Assessment techniques, tools and guide for practitioners.* ReSAKSS Working Paper 7. Washington, D.C.: Regional Strategic Analysis and Knowledge Support System facilitated by the International Food Policy Research Institute.

Berdegué, J., and G. Escobar. 2001. *Agricultural knowledge and information systems and poverty reduction.* Agricultural Knowledge and Information Systems Discussion Paper. Washington, D.C.: World Bank.

Besley, T., and A. Case. 1994. *Unnatural experiments? Estimating the incidence of endogenous policies.* NBER Working Paper 4956. Cambridge, Mass., U.S.A.: National Bureau of Economic Research.

Bhorat, H. 2007. The Nine Commandments: A South African case study identifies roles for applied research institutions. In *The Policy paradox in Africa: Strengthening links between economic research and policy-making,* ed. E. T. Ayuk and M. A. Marouani. Trenton, N.J., U.S.A., and Ottawa: Africa World Press and the International Development Research Centre.

Binswanger, H., S. Khandker, and M. Rosenzweig. 1989. *How infrastructure and financial institutions affect agricultural output and investment in India.* World Bank Working Paper 163. Washington, D.C.: World Bank.

Bolwig, S., P. Hazell, and S. Wood. 2002. Strategic assessment of land use options for Uganda—Strategic criteria for rural investment in productivity (SCRIP) phase I. Project document. International Food Policy Research Institute, Washington, D.C. Mimeo.

Booth, D., and H. Lucas. 2002. *Good practice in the development of PRSP indicators and monitoring systems.* ODI Working Paper 172. London: Overseas Development Institute.

Bourguignon, F., and L. Pereira da Silva, eds. 2003. *The impact of economic policies on poverty and income distribution: Evaluation techniques and tools.* Washington, D.C., and Oxford, U.K.: World Bank and Oxford University Press.

Bouthillier, F., and K. Shearer. 2002. Understanding knowledge management and information management: The need for an empirical perspective. *Information Research* 8 (1). <http://informationr.net/ir/8-1/paper141.html>. Accessed April 12, 2009.

Breisinger, C., X. Diao, J. Thurlow, and R. M. Al-Hassan. 2008. *Agriculture for development in Ghana: New opportunities and challenges.* IFPRI Discussion Paper 784. Washington, D.C.: International Food Policy Research Institute.

Brinkerhoff, D. 1996. Process perspectives on policy change: Highlighting implementation. *World Development* 24 (9): 1395–1401.

Byerlee, D. 2000. Targeting poverty alleviation in priority setting for agricultural research. *Food Policy* 25: 429–445.

Byerlee, D., and G. Traxler. 1996. The role of technology spillovers and economies of size in the efficiency design of agricultural research systems. Paper presented at the conference "Global Agricultural Science Policy for the 21st Century," August 26–28, Melbourne, Australia.

Cabral, L. 2006. *Poverty reduction strategies and the rural productive sectors: What have we learnt, what else do we need to ask?* Natural Resource Perspectives 100. London: Overseas Development Institute.

Canning, D., and E. Bennathan. 2000. *The social rate of return on infrastructure investments.* Policy Research Working Paper 2390. Washington, D.C.: World Bank.

Carden, F. 2009. *Knowledge to policy: Making the most of development research.* Ottawa: International Development Research Centre.

Cash, D. W. 2000. *In order to aid in diffusing useful and practical information . . . : Dross-scale boundary organizations and agricultural extension.* BCSIA Discussion Paper 2000-10. Cambridge, Mass., U.S.A.: Belfer Center for Science and International Affairs, Harvard University.

Cash, D. W., W. C. Clark, F. Alcock, N. M. Dickson, N. Eckley, D. H. Guston, J. Jager, and R. B. Mitchell. 2003. Knowledge systems for sustainable development. *Proceedings of the National Academy of Sciences* 100 (14): 8086–8091.

Chamberlin, J., J. Pender, and B. Yu. 2006. *Development domains for Ethiopia: Capturing the geographical context of smallholder development options.* Development Strategy and Governance Division Discussion Paper 43. Washington, D.C.: International Food Policy Research Institute.

Clay, E., and B. Schaffer, eds. 1984. *Room for maneuver: An exploration of public policy in agriculture and rural development.* London: Heinemann.

Colman, D., and J. Mellor. 2007. Phase I review of the Ethiopia Strategy Support Program (ESSP). Project document. International Food Policy Research Institute, Washington, D.C. Mimeo.

Court, J., and J. Young. 2003. *Bridging research and policy: Insights from 50 case studies.* Working Paper 213. London: Overseas Development Institute.

Crawford, P., and P. Bryce. 2003. Project monitoring and evaluation: A method for enhancing the efficiency and effectiveness of aid project implementation. *International Journal of Project Management* 21: 363–373.

Creech, H. 2001. *Measuring while you manage: Planning, monitoring and evaluating knowledge networks.* Winnipeg, Canada: International Institute for Sustainable Development.

Creech, H., and T. Willard. 2001. *Strategic intentions: Managing knowledge networks for sustainable development.* Winnipeg, Canada: International Institute for Sustainable Development.

Crewe, E., and J. Young. 2002. *Bridging research and policy: Context, evidence and links.* ODI Working Paper 173. London: Overseas Development Institute.

Crosby, B. 1996. Policy implementation: The organizational challenge. *World Development* 24 (9): 1403–1415.

Dalal-Clayton, B., and S. Bass. 2002. *Sustainable development strategies: A resource book.* London: Earthscan.

David, C., and K. Otsuka, eds. 1994. *Modern rice technology and income distribution.* Boulder, Colo., U.S.A.: Lynn Reinner.

Davidson, R., and J. MacKinnon. 1993. *Estimation and inference in econometrics.* London: Oxford University Press.

Deaton, A. 2009. Instruments of development: Randomization in the tropics, and the search for the elusive keys to economic development. Center for Health and Wellbeing, Research Program in Development Studies, Princeton University, Princeton, N.J., U.S.A. Mimeo.

Diao, X., and A. Nin-Pratt. 2007. Growth options and poverty reduction in Ethiopia: An economy-wide model analysis. *Food Policy* 32 (2): 205–228.

Diao, X., and D. Sarpong. 2007. *Cost implications of agricultural land degradation in Ghana: An economywide, multimarket model assessment.* IFPRI Discussion Paper 698. Washington, D.C.: International Food Policy Research Institute.

Diao, X., P. Hazell, D. Resnick, and J. Thurlow. 2007a. *The role of agriculture in development: Implications for Sub-Saharan Africa.* IFPRI Research Report 153. Washington, D.C.: International Food Policy Research Institute.

Diao, X., S. Fan, S. Kanyarukiga, and B. Yu. 2007b. *Agricultural growth and investment options for poverty reduction in Rwanda.* IFPRI Discussion Paper 689. Washington, D.C.: International Food Policy Research Institute.

Dorotinsky, B., and R. Floyd. 2004. Public expenditure accountability in Africa: Progress, lessons and challenges. In *Building state capacity in Africa: New approaches, emerging lessons,* ed. B. Levy and S. Kpundeh. Washington, D.C.: World Bank.

Dorward, A., J. Kydd, J. Morrison, and I. Urey. 2004. A policy agenda for pro-poor agricultural growth. *World Development* 32 (1): 73–89.

Douthwaite, B., T. Kuby, E. van de Fliert, and S. Schulz. 2003. Impact pathway evaluation: An approach for achieving and attributing impact in complex systems. *Agricultural Systems* 78: 243–265.

DSGD (Development Strategy and Governance Division). 2005. Ethiopia Strategy Support Program project brief. International Food Policy Research Institute, Washington, D.C. Mimeo.

———. 2006. Ghana Strategy Support Program Project Brief. International Food Policy Research Institute, Washington, D.C. Mimeo.

———. 2007a. Nigeria Agricultural Policy Support Facility: Implementation Plan. International Food Policy Research Institute, Washington, D.C. Mimeo.

———. 2007b. Terms of reference for country CAADP analysis. International Food Policy Research Institute, Washington, D.C. Mimeo.

Duflo, E., R. Glennerster, and M. Kremer. 2008. Using randomization in development economics research: A toolkit. In *Handbook of development economics,* vol. 4, ed. T. P. Schultz and J. Strauss. Amsterdam, the Netherlands: North-Holland.

Earl, S., F. Carden, and T. Smutylo. 2001. *Outcome mapping: Building learning and reflection into development programs.* Ottawa: International Development Research Centre.

Echeverria, R., ed. 1990. *Methods for diagnosing research system constraints and assessing the impact of agricultural research.* The Hague, the Netherlands: International Service for National Agricultural Research.

Evenson, R. E., and D. Gollin, eds. 2003. *Crop variety improvement and its effect on productivity: The impact of international agricultural research.* Cambridge, U.K.: CABI.

Fan, S., ed. 2008. *Public expenditure, growth and poverty in developing countries: Issues, methods and findings.* Baltimore: Johns Hopkins University Press.

Fan, S., C. Fang, and X. Zhang. 2003. How agricultural research affects urban poverty in developing countries: The case of China. *World Development* 31: 733–741.

Fan, S., P. Hazell, and S. Thorat. 1999. *Linkages between government spending and poverty in rural India.* Research Report 110. Washington, D.C.: International Food Policy Research Institute.

Fan, S., B. Omilola, and M. Lambert. 2009 Public expenditure tracking in Africa: Trends and composition. ReSAKSS Working Paper 28. Washington, D.C.: Regional Strategic Analysis and Knowledge Support System facilitated by the International Food Policy Research Institute.

Fan, S., X. Zhang, and N. Rao. 2004. *Public expenditure, growth, and poverty reduction in rural Uganda.* Development Strategy and Governance Division Discussion Paper 4. Washington, D.C.: International Food Policy Research Institute.

Fan, S., L. Zhang, and X. Zhang. 2002. *Growth, inequality and poverty in rural China: The role of public investments.* Research Report 125. Washington, D.C.: International Food Policy Research Institute.

Fan, S., M. Johnson, A. Saurkar, and T. Makombe. 2008. *Investing in African agriculture to halve poverty by 2015.* IFPRI Discussion Paper 751. Washington, D.C.: International Food Policy Research Institute.

Farrington, J., C. Thirtle, and S. Henderson. 1997. Methodologies for monitoring and evaluating agricultural and natural resources research. *Agricultural Systems* 55 (2): 273–300.

Gabre-Madhin, E. Z., and S. Haggblade. 2004. Successes in African agriculture: Results of an expert survey. *World Development* 32 (5): 745–766.

GDPRD (Global Development Platform for Rural Development). 2006. Joint donor concept on rural development. Global Development Platform for Rural Development, c/o German Federal Ministry for Economic Cooperation and Development, Bonn, Germany. Mimeo.

Gijsbers, G., W. Janssen, H. Hambly Odame, and G. Meijerink, eds. 2001. *Planning agricultural research: A sourcebook.* Wallingford, U.K.: CABI.

Gittinger, J. P. 1984. *Economic analysis of agricultural projects.* Economic Development Institute Series in Economic Development. Washington, D.C., and Baltimore: World Bank and Johns Hopkins University Press.

Greene, W. H. 1993. *Econometric analysis.* Upper Saddle River, N.J., U.S.A: Prentice Hall.

Guston, D. H. 2001. Boundary organizations in environmental policy and science: An introduction. *Science, Technology and Human Values* 26 (4): 399–408.

Haggblade, S., ed. 2004. *Building on success stories.* 2020 Focus Briefs 12. Washington, D.C.: International Food Policy Research Institute.

Hall, A., L. Mytelka, and B. Oyeyinka. 2005. *Innovation systems: Implications for agricultural policy and practice.* ILAC Brief 2. Rome: Institutional Learning and Change Initiative.

Hall, A., N. Clark, R. Sulaiman V, M. V. K. Sivamohan, and B. Yoganand. 2000. New agendas for agricultural research in developing countries: Policy analysis and institutional implications. *Knowledge, Technology, and Policy* 13 (1): 70–91.

Hoddinott, J., and E. Skoufias. 2003. *The impact of PROGRESA on food consumption.* Food Consumption and Nutrition Division Discussion Paper 150. Washington, D.C.: International Food Policy Research Institute.

Hovland, I. 2003. *Knowledge management and organizational learning: An international development perspective.* ODI Working Paper 224. London: Overseas Development Institute.

———. 2007. *Making a difference: M&E of policy research.* ODI Working Paper 281. London: Overseas Development Institute.

Howlett, M. 2002. Do networks matter? Linking policy network structure to policy outcomes: Evidence from four Canadian policy sectors, 1990–2000. *Canadian Journal of Political Science* 35 (2): 235–267.

ICASIT (International Center for Applied Studies in Information Technology). 2003. Intro to KM: Glossary of knowledge management (KM) terms. <http://www.icasit.org/km/intro/glossary.htm>. Accessed April 12, 2009.

IFAD (International Fund for Agricultural Development). 2002. *A guide for project M&E—Managing for impact in rural development.* Rome.

Jayne, T. S., M. Mukumbu, M. Chisvo, D. Tschirley, M. T. Weber, B. Zulu, R. Johannson, P. Santos, and D. Soroko. 1999. Successes and challenges of food market reform: Experiences from Kenya, Mozambique, Zambia, and Zimbabwe. MSU International Development Working Paper 72. East Lansing, Mich., U.S.A.: Michigan State University, Department of Agricultural Economics.

Johnson, M., and D. Resnick. 2004. *Strategic analysis and knowledge support systems for rural development strategies in Sub-Saharan Africa.* Development Strategy and Governance Division Discussion Paper 14. Washington, D.C.: International Food Policy Research Institute.

Johnson, M., S. Wood, J. Hoddinott, P. Hazell, and J. Lamb. 2003. Strategic analysis and knowledge support system (SAKSS) for small farmer led agricultural growth. Background paper prepared for the United States Agency for International Development. International Food Policy Research Institute, Washington, D.C. Mimeo.

Johnson, M., R. Birner, J. Chamberlin, X. Diao, S. Fan, A. Nin Pratt, D. Resnick, L. You, and B. Yu. 2008. *Regional strategic alternatives for agriculture-led growth and poverty reduction in West Africa.* ReSAKSS Working Paper 22. Washington, D.C.: Regional Strategic Analysis and Knowledge Support System facilitated by the International Food Policy Research Institute.

Jones, N., A. Datta, and H. Jones, with Evidence-Based Policy in Development Network partners. 2009. *Knowledge, policy and power: Six dimensions of the knowledge–development policy interface.* London: Overseas Development Institute.

Juma, C., and N. Clark. 1995. Policy research in Sub-Saharan Africa: An exploration. *Public Administration and Development* 15: 121–137.

Keeley, J., and I. Scoones. 1999. *Understanding environmental policy processes: A review.* IDS Working Paper 89. Sussex, U.K.: Institute of Development Studies.

Kessides, C. 1993. *The contributions of infrastructure to economic development: A review of experience and policy implications.* Discussion Paper 213. Washington, D.C.: World Bank.

Killick, T. 1976. The possibilities of development planning. *Oxford Economic Papers* 28 (2): 161–184.

Kiregyera, B. 2008. *Reforming national statistical systems in Africa: A case for making the national statistics offices autonomous.* Addis Ababa, Ethiopia: United Nations Economic Commission for Africa, African Centre for Statistics.

Kusek, J. Z., and R. C. Rist. 2004. *Ten steps to a results-based monitoring and evaluation system: A handbook for development practitioners.* Washington, D.C.: World Bank.

Livny, E., A. Mehendale, and A. Vanags. 2006. Bridging the research and policy gaps in developing and transition countries—Analytical lessons and proposals for action. A synthesis of findings from the Global Development Network's Bridging Research and Policy Project. World Bank, Global Development Network, Washington, D.C. Mimeo.

Mackay, K. 2007. *How to build M&E systems to support better government.* Washington, D.C.: World Bank.

Maredia, M. K. 2009. Improving the proof—Evolution of and emerging trends in impact assessment: Methods and approaches in agricultural development. Background paper prepared for the Millions Fed Project. International Food Policy Research Institute, Washington, D.C. Mimeo.

Masters, W., T. Bedingar, and J. Oehmke. 1998. The impact of agricultural research in Africa: Aggregate and case study evidence. *Agricultural Economics* 19: 81–86.

Mendizabal, E. 2006. *Understanding networks: The functions of research policy networks.* ODI Working Paper 271. London: Overseas Development Institute.

Minot, N., and B. Baulch. 2005. Poverty mapping with aggregate census data: What is the loss in precision? *Review of Development Economics* 9 (1): 5–24.

Montjoy, R., and L. O'Toole. 1979. Toward a theory of policy implementation: An organizational perspective. *Public Administration Review* 39 (5): 465–476.

Neilson, S. 2001. Knowledge utilization and public policy processes: A literature review. Background paper. Evaluation Unit, International Development Research Centre, Ottawa. Mimeo.

NEPAD (New Partnership for Africa's Development). 2003. Comprehensive Africa Agriculture Development Programme (CAADP). Midrand, South Africa. Mimeo.

ODI (Overseas Development Institute). 2003. *PRSP monitoring in Africa.* PRSP Synthesis Note 7. London.

Omamo, S. W. 2004. *Bridging research, policy, and practice in African agriculture.* Development Strategy and Governance Division Discussion Paper 10. Washington, D.C.: International Food Policy Research Institute.

Omamo, S., X. Diao, S. Wood, J. Chamberlin, L. You, S. Benin, U. Wood-Sichra, and A. Tatwangire. 2006. *Strategic priorities for agricultural development in Eastern and Central Africa.* IFPRI Research Report 150. Washington, D.C.: International Food Policy Research Institute.

Pender, J., P. Jagger, E. Nkonya, and D. Sserunkuuma. 2001. *Development pathways and land management in Uganda: Causes and implications.* EPTD Discussion Paper 85. Washington, D.C.: International Food Policy Research Institute.

Perkin, E., and J. Court. 2005. *Networks and policy processes in international development: A literature review.* ODI Working Paper 252. London: Overseas Development Institute.

Pitt, M., M. Rosenzweig, and D. Gibbons. 1993. The determinants and consequences of the placement of government programs in Indonesia. *World Bank Review* 7 (3): 319–348.

Ravallion, M. 2008. *Evaluation in the practice of development.* Policy Research Working Paper 4547. Washington, D.C.: World Bank.

———. 2009. Should the randomistas rule? *The Economists' Voice* 6 (2): 1–5

Resnick, D., and R. Birner. 2008. *Agricultural strategy development in West Africa: The false promise of participation?* IFPRI Discussion Paper 844. Washington, D.C.: International Food Policy Research Institute.

Rosegrant, M. W., M. S. Paisner, S. Meijer, and J. Witcover. 2001. *2020 global food outlook: Trends, alternatives, and choices.* IFPRI Food Policy Report. Washington, D.C.: International Food Policy Research Institute.

Rosegrant, M., C. Ringler, S. Msangi, T. Sulser, T. Zhu, and S. Cline. 2008. International Model for Policy Analysis of Agricultural Commodities and Trade (IMPACT): Model description. International Food Policy Research Institute, Washington, D.C. Mimeo.

Rossi, P. H., and H. E. Freeman. 1993. *Evaluation: A systematic approach,* 5th ed. Newbury Park, Calif., U.S.A.: Sage.

Ryan, J .G. 1999. *Assessing the impact of policy research and capacity building by IFPRI in Malawi.* Impact Assessment Discussion Paper 11. Washington, D.C.: International Food Policy Research Institute.

Ryan, J. G., and J. L. Garrett. 2003. *The impact of economic policy research: Lessons on attribution and evaluation from IFPRI.* Impact Assessment Discussion Paper 20. Washington, D.C.: International Food Policy Research Institute.

Sabatier, P., ed. 2007. *Theories of the policy process,* 2nd ed. Boulder, Colo., U.S.A.: Westview Press.

Sartorius, R. 1996. The third generation logical framework approach: Dynamic management for agriculture research projects. *Journal of Agriculture Education and Extension* 2 (4): 49–62.

Schiffer, E. 2008. *Tracing power and influence in networks: Net-Map as a tool for research and strategic network planning.* IFPRI Discussion Paper 772. Washington, D.C.: International Food Policy Research Institute.

Scott, C. 2005. Measuring up to the measurement problem: The role of statistics in evidence based policy-making. Paper prepared for the PARIS21 Partnership in Statistics for development in the 21st Century, London School of Economics. Mimeo.

Smutylo, T. 2005. *Outcome mapping: A method for tracking behavioral changes in development programs.* ILAC Brief 7. Rome: Institutional Learning and Change Initiative.

Spielman, D., and R. Pandya-Lorch, eds. 2009. *Millions fed: Proven successes in agricultural development.* Washington, D.C.: International Food Policy Research Institute.

Springer-Heinze, A., F. Hartwich, J. Henderson, D. Horton, and I. Minde. 2003. Impact pathway analysis: An approach to strengthening the impact orientation of agricultural research. *Agricultural Systems* 78: 267–285.

Start, D., and I. Hovland. 2004. *Tools for policy impact: A handbook for researchers.* London: Overseas Development Institute.

Stone, D. 2002. The dilemmas of "bridging research and policy." *Compare* 32 (3): 285–296.

Stone, D., S. Maxwell, and M. Keating. 2001. Bridging research and policy. Radcliffe House, University of Warwick, Sussex, U.K. Mimeo.

Tadesse, M. 2006. A landscape of knowledge systems in Ethiopia: The case of public institutions involved in rural development. Paper presented at the Ethiopia Strategy Support Program Conference, June 6–8, Addis Ababa, Ethiopia.

Thomas, J. W., and M. S. Grindle. 1990. After the decision: Implementing policy reforms in developing countries. *World Development* 18 (8): 1163–1181.

Thurlow, J. 2008. *Agricultural growth and investment options for poverty reduction in Mozambique.* ReSAKSS Working Paper 20. Washington, D.C.: Regional Strategic Analysis and Knowledge Support System facilitated by the International Food Policy Research Institute.

Thurlow, J., S. Benin, X. Diao, A. Kebba, and N. Ofwono. 2008a. *Agricultural growth and investment options for poverty reduction in Uganda.* IFPRI Discussion Paper 790. Washington, D.C.: International Food Policy Research Institute.

Thurlow, J., S. Benin, X. Diao, H. Kalinda, and T. Kalinda. 2008b. *Agricultural growth and investment options for poverty reduction in Zambia.* IFPRI Discussion Paper 791. Washington, D.C.: International Food Policy Research Institute.

Timmer, C. P. 1998. *Adding value through policy-oriented research: Reflections of a scholar-practitioner.* Impact Assessment Discussion Paper 4. Washington, D.C.: International Food Policy Research Institute.

Tschirley, D., and M. Mathenge. 2003. Developing income proxy models for use by the USAID mission in Kenya: A technical report. Working Paper 7. Nairobi, Kenya: Egerton University, Tegemeo Institute of Agricultural Policy and Development.

Van de Walle, D. 1998. Assessing welfare impacts of public spending. *World Development* 26 (3): 365–379.

Vogel, S. J. 1994. Structural changes in agriculture: production linkages and agricultural demand-led industrialization. *Oxford Economic Papers* 46 (1): 136–156.

Wangwe, S. 2005. A review of how external agencies can influence research agendas and research policy links. Economic and Social Research Foundation, Dar-es-Salaam, Tanzania. Mimeo.

White, H. 2006. *Impact evaluation: The experience of the Independent Evaluation Group of the World Bank.* Munich Personal Research Papers in Economics (RePEc) Archive Working Paper 1111. Washington, D.C.: World Bank.

Wiig, K. 1997. Knowledge management: Where did it come from and where will it go? *Expert Systems with Applications* 13 (1): 1–14.

Wildavsky, A. 1973. If planning is everything, maybe it's nothing. *Policy Sciences* 4: 127–153.

Wingfield-Digby, P. 2007. *Towards reforming national statistical agencies and systems: A survey of best practice countries with effective national statistical systems in Africa.* ACBF Best Practice Study Series. Harare, Zimbabwe: African Capacity Building Foundation.

Wood, S., and J. Chamberlin. 2003. Enhancing the role of spatial analysis in strategic impact assessment: Improving data resolution for regional studies. *Quarterly Journal of International Agriculture* 42 (2): 167–187.

Wood, S., K. Sebastian, F. Nachtergaele, D. Nielsen, and A. Dai. 1999. *Spatial aspects of the design and targeting of agricultural development strategies.* EPTD Discussion Paper 44. Washington, D.C.: International Food Policy Research Institute.

Wood, S., L. You, and W. Baitx. 2001. Dynamic Research EvaluAtion for Management (DREAM). Washington, D.C.: International Food Policy Research Institute.

World Bank. 1994. *World development report 1994: Infrastructure.* Washington, D.C.

———. 2003. *A user's guide to poverty and social impact analysis.* Washington, D.C.

———. 2004a. *Monitoring and evaluation: Some tools, methods and approaches.* Washington, D.C.

———. 2004b. *The Poverty Reduction Strategy Initiative: An independent evaluation of the World Bank's support through 2003.* Washington, D.C.

———. 2007a. *Tools for institutional, political, and social analysis of policy reform: A sourcebook for development practitioners.* Washington, D.C.

———. 2007b. *World development report 2008: Agriculture for development.* Washington, D.C.

Yeh, A. 1999. Decision support with geographic information systems. In *Decision support systems for sustainable development,* ed. G. E. Kersten, Z. Mikolajuk, and A. Yeh. Boston: Kluwer Academic.

You, L., S. Wood, U. Wood-Sichra, and J. Chamberlin. 2007. Generating plausible crop distribution maps for Sub-Saharan Africa using a spatial allocation model. *Information Development* 23 (2–3): 151–159.

Young, J. 2005. Research, policy and practice: Why developing countries are different. *Journal of International Development* 17: 727–734.

About the Authors

Michael Johnson is a research fellow in the Development Strategy and Governance Division of the International Food Policy Research Institute, Washington, D.C.

Kathleen Flaherty is a data analysis specialist in the Environment and Production Technology Division of the International Food Policy Research Institute, Washington, D.C., and is based in Rome.

PHOTO CREDITS:

Page xvi © 2009 Elias Zerifu/IFPRI
Page 12 (left) © 2008 Mebratie Mihretu
Page 12 (right) © 2009 Elias Zerifu/IFPRI
Page 14 © 2008 Mebratie Mihretu
Page 34 © 2009 Elias Zerifu/IFPRI
Page 44 (left) © 2008 Ekin Birol/IFPRI
Page 44 (center) © 2010 Charles Gamde/IFPRI
Page 44 (right) © 2006 Svetlana Edmeades
Page 46 © 2008 Ekin Birol/IFPRI
Page 56 © 2010 Charles Gamde/IFPRI
Page 68 © 2006 Svetlana Edmeades